After the Rescue

CONTEMPORARY ANTHROPOLOGY OF RELIGION,

A series published with the Society for the Anthropology of Religion

Robert Hefner, Series Editor
Boston University

Published by Palgrave Macmillan

Body / Meaning / Healing
By Thomas J. Csordas

The Weight of the Past: Living with History in Mahajanga, Madagascar
By Michael Lambek

After the Rescue

Jewish Identity and Community in Contemporary Denmark

Andrew Buckser

palgrave
macmillan

AFTER THE RESCUE
Copyright © Andrew Buckser, 2003.
All rights reserved. No part of this book may be used or reproduced in
any manner whatsoever without written permission except in the case of
brief quotations embodied in critical articles or reviews.

First published 2003 by
PALGRAVE MACMILLAN™
175 Fifth Avenue, New York, N.Y. 10010 and
Houndmills, Basingstoke, Hampshire, England RG21 6XS.
Companies and representatives throughout the world.

PALGRAVE MACMILLAN is the global academic imprint of the
Palgrave Macmillan division of St. Martin's Press, LLC and of Palgrave
Macmillan Ltd. Macmillan® is a registered trademark in the United States,
United Kingdom and other countries. Palgrave is a registered trademark in
the European Union and other countries.

ISBN 0–312–23945–9 hardback
ISBN 1–4039–6270–7 paperback

Library of Congress Cataloging-in-Publication Data is available from the
Library of Congress.

A catalogue record for this book is available from the British Library.

Design by Letra Libre, Inc.

First edition: May 2003
10 9 8 7 6 5 4 3 2 1

Printed in the United States of America.
Transferred to digital printing in 2007.

For Sue
The girl that I adore

Contents

Acknowledgments

This study is based on fieldwork funded by grants from the American-Scandinavian Foundation, the Lois Roth-Thomsen Endowment, the Purdue Research Foundation, the Purdue University School of Liberal Arts, and the Jewish Studies Program at Purdue University. In addition, the Institute for European Ethnology and the Department of Anthropology at the University of Copenhagen both allowed me to use their office space and facilities during parts of my research. I am grateful to all of these institutions for their support.

Like most anthropologists, I am most grateful of all to the people with whom I have done my fieldwork, who have put up with my nosiness and foolish questions for more than five years now. Space does not permit me to mention more than a handful by name, but I hope that all of the Copenhagen Jews with whom I have spoken will accept my deepest thanks for their time, their kindness, and their invariably thoughtful perspectives on Jewish life. I would especially like to express my gratitude to Dan Besjakov, without whom this study would have been impossible; to Karen Lisa Salamon, for her hospitality, her anthropological insights, and her extraordinary patience; to Irina Magieres, a wonderful and inspiring guide to the Polish Jewish community; and to the irrepressible David Gerschwald. A number of community leaders were very generous with their time and advice; special thanks are due to Bent Lexner, Bent Melchior, Jacques Blum, Erik Guttermann, and Yitzchok and Rochel Loewenthal. I would also like to thank Simon and Dina Bergmann, Anne Boukris, Grethe Gerschwald, Linda Hertzberg and Bjarne Karpantschof, the Königshøfer family, Sara Kviat, Tatjana Lichtenstein, the Meyerowitsch family, the Nathan family, Herbert Pundik, the Saarde family, Martin Sala-mon, Torben Samson, and the staffs of the Mosaiske Troessamfund, the Carolineskole, and the Gan Aviv børnehave.

A number of Danish scholars have been instrumental to this project, first and foremost Palle Ove Christiansen of Copenhagen University. Professor Christiansen has been unstinting in his advice and his hospitality throughout all my research in Denmark, and I am grateful for his friendship and support. I would also like to thank Bent Blüdnikow, Merete Christiansen, Lasse Dencik, Ulf Haxen, Karen Oslund, Jonathan Schwartz, Hanne Trautner-Kromann, and Mikkel Venborg Pedersen. Thanks also, as always, to Barbara Lehman of the Danish Fulbright Commission.

During my first visits for this project, I was allowed to take part in a series of lectures and seminars organized by the Thanks to Scandinavia Foundation. My thanks to the foundation; to its director, Judy Goldstein; and to the late and much-missed Dan Cooperman.

In the United States, many scholars have given advice and criticism on my writings about Danish Jewry, and in doing so they have contributed greatly to this study. I would like to particularly thank Steve Glazier, Carol Greenhouse, Herbert Lewis, Charles Lindholm, Brian Palmer, Leonard Plotnicov, Benson Saler, and Susan Sered. Most of all, I would like to thank the late Morton Klass, a model not only of committed scholarship but of a life well lived.

Many thanks to my editors at Palgrave, Kristi Long and Roee Raz, both for their expertise and for their patience.

Throughout this project, I have worked as a professor in the Department of Sociology and Anthropology at Purdue University. I am grateful to the department as a whole and its members as individuals for their support and encouragement during the process. My department heads, Dean Knudsen and Carolyn Perrucci, did everything in their power to assist me in getting the time and funding necessary for my work. My colleagues, likewise, offered abundant help and advice whenever it was needed. I would particularly like to thank Marty Patchen, Rachel Einwohner, Michael Watson, Rich Blanton, Jim Davidson, Myrdene Anderson, and Ginger Hofman. The departmental staff has provided wonderful support; the efforts of Marcy Jasmund, Barb Puetz, Phyllis Hunt, Dianne Livingston, Michael Copeland, Pat Henady, Evelyn Douthit, Ruth Ann Brooks, and Dawn Stahura are especially appreciated. I also owe a special debt of gratitude to the Purdue Jewish Studies Program, where I was able to present some early versions of the analysis in this book. Joe Haberer and

Bob Melson were particularly generous with their time and insight, and they have contributed greatly to this study.

Finally, I would like to thank my family, for their support and forbearance over the course of my fieldwork and writing. My mother, Carol Dolan, gave me not only encouragement but also expert editing and criticism; my parents-in-law, Harriet and Irwin Rofman, provided research and an insider's perspective on many of the subjects covered here. Lisbeth Svalgaard, Gerda Jeppesen, Lis Christensen, and Ellen and Ole Tarri gave me a home away from home on my visits back to Jutland. My dear children, Rachel, William, Sarah, Edward, and little James, endured my long absences in the field and gave me the best reason in the world to come back. Most of all, I would like to thank my wife, Susan Ann Buckser, for her advice, her encouragement, and her love over a long and often trying project. It is to her that I dedicate this book, and my heart.

Introduction

I make this covenant, and I make it with you: you shall be the father of a host of nations. Your name shall no longer be Abram, your name shall be Abraham, for I make you the father of a host of nations. I will make you exceedingly fruitful; I will make nations out of you, and kings shall spring from you. I will fulfill my covenant between myself and you and your descendents after you, generation after generation, an everlasting covenant, to be your God, yours and your descendents' after you. As an everlasting possession I will give you and your descendents after you the land in which you are now aliens, all the land of Canaan, and I will be God to your descendents.

Genesis 17:4 8

A long time ago, according to Jewish myth, the Creator of the Universe made a bargain with the shepherd Abram. If Abram would take the Creator as his God, worship only Him and obey His commandments, the Creator promised to give Abram a land of his own. It would be a beautiful land, the land of Zion, flowing with milk and honey and fertility, where the shepherd's descendents could flourish in countless numbers. Abram accepted the bargain, and under his new name of Abraham he founded the lineage that would become the Jews. His descendents flourished in Zion for centuries, according to the myth; even when they were driven out of it, scattered to the four winds at the dawn of the Christian era, the image of Zion bound them together as a people. The memory of the promised land, and the prophecies of an eventual return to it, stood at the center of the Jewish religious and cultural world throughout the long centuries of the Diaspora.

Six thousand years or so later, the Jews were offered another bargain—not by God this time, but by the secularizing nation-states fast displacing Him from the center of European culture. This bargain called on the Jews to give up the autonomy they had developed over the centuries of exile, the special status and community institutions that distinguished them from their non-Jewish neighbors. In return, they would receive acceptance, full membership in the local societies that for centuries had ostracized and repressed them. In changing Jews from outcasts to citizens, the Enlightenment offered a new promised land, one free of the prejudice and discrimination that had shackled the Jews since their arrival in Europe.

One can debate whether God kept the promise he made to Abram. While the Jews did settle in Zion, they did not stay there; again and again, foreign conquerors overran the land and sent the Jews into exile, until Jewishness became synonymous with wandering and many regarded the promised land as a fantasy. The meaning of the promised land remains hotly contested as I write this, both among theologians and increasingly among rival armies. Of the second promise, on the other hand, there is little room for debate. European politicians did indeed abolish most of the Jews' distinctive institutions, but the promised equality of full citizenship never really materialized. In most of Europe, anti-Semitism remained ingrained and vicious, no matter how fervently Jews embraced their various new national cultures. Whatever their legal status, Jews remained outsiders as long as they held to their faith, and often even when they forsook it. The hollowness of the promise emerged with horrific clarity in the middle of the twentieth century, when what had been one of the most progressive of the liberal states launched a merciless assault on European Jewry. The Holocaust destroyed the facade of liberalism even as it slaughtered the Jews of Europe, exposing the emptiness of the glittering rhetoric of the leaders of the Enlightenment.

There was in this, however, arguably a single exception: Denmark, a small and otherwise unremarkable country on the southern edge of Scandinavia. Denmark was not a paradise for Jews, and many forms of discrimination survived the establishment of Jewish citizenship in 1814. Still, the experience of Danish Jewry in the nineteenth century generally followed the predictions of the architects of emancipation. Jews became steadily more accepted as they immersed themselves in Danish culture; their living standards rose, some became leaders of science and industry, and the anti-Semitism of the larger society grew steadily less pernicious. By the early twentieth century, Danish Jewry had become an integral element of the national middle class. And as the Holocaust raged around Europe in the 1940s, something happened in Denmark that can be plausibly described by theologians, if not by anthropologists, as a miracle. The Danes refused to join the parade of nations that abandoned or actively persecuted Jews during the German occupation. When the Germans decided to eliminate the Danish Jews anyway in 1943, the nation's population rose with astonishing unanimity to defend them. Alone in occupied Europe, Denmark saved almost all of its Jews from the Holocaust, and in the years afterward Denmark blotted anti-Semitism almost completely from its culture. The result was a country that could lay plausible claim to having fulfilled the liberal vision of Jewish citizenship set out at the close of the eighteenth century. While no place in Europe has fully kept the promise made to Jews by the Enlightenment, Denmark has perhaps come closer than any other.

This book explores Jewish life in this other promised land. It considers how Jewish society has developed in a setting that lacks the exclusion so basic to the Jewish experience in most of the Diaspora. It asks whether and how Jewish identity persists in the absence of the social walls that defined it for so long. As Jewishness becomes a chosen identity, and as it coexists with non-religious national and regional identities, what becomes of Jewish community and tradition? Can the different elements of the Jewish tradition—its institutions, its beliefs, its folklore, and so on—survive outside of a milieu that forcibly binds them together? Will they simply dissolve, withering under the onslaught of secularizing modernity? Or will they change into something new, finding new forms of relevance and meaning in an open society? If the latter, what will these forms be, and how will they be influenced by the particular character of the larger national culture? This study examines the interaction of Danish and Jewish cultures in contemporary Copenhagen and how each helps shape the other.

The answers to these questions have implications not only for the study of European Jewry but for anthropology and the social sciences as a whole. They echo questions that have increasingly troubled scholars of other ethnic cultures and indeed of modern culture and society generally. For the breakdown of the walls that separated Jews from their neighbors during emancipation was part of a larger cultural process, one in which states throughout the West undermined the autonomy of the small-scale organizations and communities within their borders. Modernization has steadily eroded both the legal and the cultural integrity of the local associations that once defined most of human existence; neighborhoods, kin groups, and occupational affiliations have lost much of their relevance, while large-scale government and economic institutions have become ever more powerful. For many scholars, this process raises questions for all communities very similar to those that concern the Jews. How do communities cohere in the ever more fragmented and globalized world of late modernity? As individual lives throughout the West become increasingly dominated by state and corporate bureaucracies, as modernizing processes break up parochial identities, how do small-scale affiliations retain any strength and meaning at all? Has the notion of community itself become outdated, an expression of romantic nostalgia rather than a meaningful social force? If not, what is the meaning that it retains? While this book focuses on a small and little-known population, its implications extend to communities throughout the modern world.

Understanding that response, I will argue, requires some rethinking of our standard ideas about what communities are and how they work. The contemporary Jewish community of Denmark presents a seeming contradiction, in that it combines a manifest institutional integrity with an equally evident fragmentation among its members. The Copenhagen Jews comprise, on the one hand, one of the most active and engaged religious groups in Scandinavia; their few thousand members operate an impressive set of religious, social, and cultural institutions. Jews generally report a strong consciousness of Jewish identity and a serious interest in the affairs of the Jewish world. At the same time, however, Copenhagen Jewry represents an exceptionally fractious and fragmented group of people. Jews disagree about virtually every element of Jewish life, ranging from the boundaries of the community to the conduct of ritual practice to the operation of Jewish social institutions. Community politics involves a host of cross-cutting parties and interest groups, many of which oppose each other with intense bitterness. No consensus exists on the basic issues that define group identity; asking what a Jew is, for example, can produce a variety of responses, each based on a deeply held and fiercely defended understanding of who counts. Nor do Jews share views on the relationship between Danish and Jewish identities, or on the significance of kosher practice or the state of Israel. Most of the Jews I interviewed worried about this fragmentation, and many felt that the community was close to falling apart. Despite some clear signs that the community is thriving, then, the sense of common identity and shared worldviews usually associated with community has little manifestation in Jewish Denmark. Indeed, some people, including some Copenhagen Jews, argue that no Jewish community exists in Denmark at all.

I think that they are wrong. There is a Jewish community, one with a profound meaning and significance for those affiliated with it. That community is not, however, the sort of unified *gemeinschaft* imagined by classical sociological community studies or by popular films about the Lower East Side. The tight Jewish communities of *Life Is with People* or *Crossing Delancey* have no parallel in Copenhagen; such a group, united by a common worldview and sharply distinguished from its neighbors, could never survive in a setting like contemporary Denmark. This Jewish community is best understood rather as a sort of toolbox, a set of symbolic resources on which certain individuals can draw to construct definite notions of self in the increasingly fluid world of late modernity. Following Anthony Cohen (1985), I argue in this book that community consists not so much in a group of people as in a body of symbols, a set of references that individuals can construe in many different ways. What makes people part of a community is not their agreement on the meaning of particular symbols, but their use of a common symbolic framework to construct their understandings of self and world. This approach allows us to understand the persistent appeal of ethnic communities in the fragmented world of late modern society. Such communities offer a sense of rootedness and authenticity hard to come by in places like Denmark and the United States. The hunger of individuals for these roots suggests that Jewish communities can survive, and indeed become essential, even in this very different sort of promised land.

In the chapters to follow, we will trace the various forms in which the Jewish tradition shapes and informs the lives of Jews in contemporary Copenhagen. We will consider Jewishness as a religious identity, a social identity, an institutional affiliation, and more. Each of these perspectives invokes Judaism in a somewhat different way, implying different questions about the nature of Jewishness and different cleavages among community members. Each, moreover, can only be understood in the context of the larger Danish cultural environment. The details of this interaction are often complex, and they can lead to unexpected results. This study therefore takes an ethnographic approach, training a close eye on the details of the personal and institutional interactions through which Jewishness in Denmark is enacted. We will try to explore not only grand themes and large-scale patterns of the Jewish world but more importantly the concrete and personal structures in which everyday Jewish life is lived.

This is not the first book to take such an approach to a Jewish community. While ethnographies of Jews were relatively rare for much of the twentieth century, the past several decades have seen a rapidly growing anthropological interest in the subject. Much of this work has focused on Jews in the United States. Studies like Myerhoff (1978), Kugelmass (1986), Furman (1987), and Shokeid (1988) offer comprehensive analyses of particular Jewish communities in a variety of settings; a number of shorter works have focused on specific questions about individual Jewish groups (see, for example, those collected in Belcove-Shalin 1995, Zenner 1988, Goldberg 1987, Zuckerman 1999, and Kugelmass 1988). Readers will note many parallels between their descriptions and the one presented here. Studies of Jewish communities have also been undertaken elsewhere in Europe, including Boyarin's (1991) on Paris and Berghahn's (1984) on London (see also Webber's excellent collection [1994a], as well as Bunzl 2000, Mars 1999, Goluboff 2001, and Hofman 2000). Anthropologists working in Israel have produced some of the most intriguing ethnographies, exploring the rich possibilities there for the interaction and confrontation of different variants of Jewish ethnicity and religion (see, for example, Sered 1988, 1992, Seeman 1999, Golden 2001, Torstrick 2000, and Zerubavel 1994). Some anthropologists, meanwhile, have focused on specific aspects of Jewish modernity that transcend specific local settings. The works of Jonathan Boyarin (Boyarin 1992; Boyarin 1996), Alain Finkielkraut (1994), and Walter Zenner (1991; 1985), among others, offer insights that illuminate conditions among Jews in Denmark as well as the rest of the West.

There are a variety of reasons for the new interest of anthropology in the Jews, one of which is a growing interest in ethnic diasporas more generally (see Kearney 1995; Malkki 1995; Shukla 2001; Yelvington 2001). Diasporas make interesting subjects for anthropologists, because they foreground the questions of cultural continuity, definition, and change that have animated much of recent anthropological theory. Postmodern anthropology has focused heavily on the conflict over cultural forms that goes on in all societies; increasingly, anthropologists have shown that traditions and customs are not simply accepted and passed down in societies but are actively created and re-created through an often contentious political process. This process emerges with unusual clarity in diasporic groups, who must constantly wrestle with the definition of group boundaries and the meaning of cultural traditions. Many of the questions that animate studies of the Danish Jews are also reflected in work on the Chinese (e.g., Louie 2000; Nonini 1999), African (e.g., Glazier 1996; Lake 1995; Murphy 1995; Pitts 1996; Yelvington 2001), South Asian (e.g., Bhatt 2000; Raj 2000; Shukla 2001), and other diasporas.

This study is intended in part as a contribution to this larger literature on transnational diaspora societies. At first glance, its subject may not seem likely to add much to the discussion. The Danish Jews are, after all, a small group with little influence on the larger world. They number perhaps one fiftieth of the Jewish population in England, and less than one thousandth of that in America. Even in Denmark, they are all but invisible to the larger population. When I tell people in America about my work, they usually react not with riveted interest, but with a rather quizzical "You mean there are Jews in Denmark?" The size of the community, however, and the country is rather deceptive, for the implications of their experience are profound. The Jews of Denmark find themselves in a situation almost unknown to history, even in the United States: a situation in which Jews have been embraced by a Christian country, in which anti-Semitism has ebbed almost to insignificance, in which the glittering promises of acceptance so often made to society's outsiders have, one might argue, been kept. The outcome of this situation may tell us something important about the possibilities for other outsiders, Jewish or not. As the Danish Jews make sense of their position, as they struggle to unite the traditions of the Jewish past and the culture of the Danish present, they suggest some of the limits, the possibilities, and the irreducible dilemmas of the acceptance to which so many of those on the margins of society aspire.

Studying the Jews of Copenhagen

Almost all of the Jews in Denmark live in Copenhagen, a prosperous city of a million and a half people at the western edge of the Baltic Sea. This fact could easily escape a casual visitor, since there are so few signs of a Jewish presence in the city. In contrast to the Muslim population, Jews do not live in any particular quarter or neighborhoods, nor do they have any style of dress or behavior that visibly distinguishes them from other Danes. Judaism has little visibility in public culture; grocers do not stock kosher food, for example, nor do Jews figure prominently in films or television programs. No distinctively Jewish occupations exist in Denmark, and the economic characteristics of Jews differ little from those of other Danes. Jewish institutions and social organizations do exist, but they neither advertise their presence nor welcome visitors. A person could easily live for years in Copenhagen without realizing that Jews live there; I suspect that many do.

Jews do live there, of course, and they have for quite some time. They first arrived in the city in the seventeenth century, and the formal Jewish

Community dates to 1684. Today the Jews constitute the most established ethnic and religious minority group in the country. They maintain an impressive institutional structure, including an imposing main synagogue, two rival synagogues, a school, two nursing homes, and a wide variety of social, cultural, and political organizations. A number of them have become prominent figures in Danish culture over the years, from writers like Georg Brandes and Meier Goldschmidt to industrialists like David Baruch Adler and Meyer and Jacob Bing. Victor Borge—born Børge Rosenbaum—became perhaps the world's best-known Dane in the second half of the twentieth century.[1] For so small a group, the Jews also display a striking amount of internal diversity. The waves of immigration that have brought Jews to Denmark—from Germany in the eighteenth and nineteenth centuries, from Eastern Europe at the opening of the twentieth, from Poland in 1970, and from the Middle East in the 1990s—have left enduring cleavages within the Jewish community. Other divisions relate to disputes over religious practice, political affiliations, family rivalries, and administrative disagreements. In some cases, these differences have led to the development of alternative institutions. A disagreement in 1913, for example, led a conservative faction to establish an alternative synagogue and a network of voluntary organizations. More commonly, the differences manifest themselves in forms of everyday religious practice and cultural engagement. Danish Jewry includes people who observe Jewish dietary laws meticulously and people who serve roast pork in front of Christmas trees; people who spend six months of each year in Israel and people who identify completely as Danes; people who say prayers daily in their homes and people who set foot in a synagogue only for weddings and bar mitzvahs. In their understandings of what Jewishness is, and its place in everyday behavior, the Danish Jews encompass virtually every point in the Jewish ideological spectrum.

This diversity implies some practical difficulties for an anthropological study of Danish Jewry. Some of the basic questions that structure the process of field research—Where is the community? Who belongs to it? What kinds of behaviors are relevant to its existence?—have no obvious or objective answers. Many, indeed, are the subjects of vigorous dispute among the Jews themselves. The size of the Jewish community, for example, is the subject of serious differences of opinion. How many Jews live in Copenhagen? That depends on how one defines a Jew. Does one include people who have Jewish parents but never participate in Jewish activities? Does one include children of mixed marriages, who may regard themselves as Jewish even if the rabbi does not? What about people who have severed their ties with the formal Jewish Community? Any answer assumes a particular definition of Jewish identity, and consequently different Jews offer different estimates of the community's size. In interviews, I heard estimates ranging from 2,500 to 12,000, each with a plausible justification.[2] As a practical matter, such uncertainties preclude any "natural" structure either for the fieldsite or for the process of fieldwork. The anthropologist must make a conscious decision about what sort of Jewish community he or she intends to study, and then shape the field methods to provide access to that population.

In this study, I have defined Jewishness in a fairly broad manner. Since I am interested particularly in issues of identity, I have viewed all those as Jewish who regard themselves as such; this includes nonpracticing Jews and children of mixed marriages as much as the inner circle of the Great Synagogue. My aim has been to explore the wide variety of ways in which Jewishness is understood, experienced, and enacted, regardless of the legal or halakhic status of the people involved. Consequently, I have designed my fieldwork to reach as wide a spectrum of Danish Jewry as possible. I have worked with both orthodox and liberal Jews, descendents of the early German-Jewish settlers and recent immigrants to Denmark, the leaders of the Jewish Community and some of their most bitter enemies, synagogue stalwarts as well as avowed atheists. Not all Danish Jews would draw the boundaries of the community as broadly as I have here, and some would draw them much more narrowly indeed. Still, I think that the picture that emerges would be a familiar one to most of them. Danish Jews tend to think of Jewishness in primordial terms, as something deeper and more innate than a simple social label. Despite their conflicts over the abstract nature of Jewishness, they tend to agree on who is and is not part of the Jewish world; never did I interview a person claiming to be a Jew, then later have that specific person's identity disputed by another informant. In practice, even those who conceptualize the community most narrowly tend to relate to it in a fairly inclusive way, and my approach here follows that tendency.

The fieldwork for this study took place in a series of visits to Copenhagen between 1996 and 2001. Visits ranged in length from two to six weeks, mainly in the summer but occasionally at other times. Much of the data derives from interviews, both structured and unstructured. Over the course of the fieldwork, I interviewed about 120 Jews in a variety of settings, almost always in Danish, conducting multiple interviews when possible.[3] In addition, I carried out participant observation in a variety of

Jewish activities. These included regular religious services at all of the synagogues, as well as holiday and life-cycle rituals at the Great Synagogue. I took part in meetings of social organizations, recreational groups, and institutions like the Jewish school and the Chabad House. I visited Jews in their homes whenever possible, and during one visit I lived with a Jewish family. This sort of fieldwork differs from the long-term immersion of the classic anthropological method, the approach I had taken to my earlier research in 1990–1991 (see Buckser 1996a). A series of short visits does not allow for the same level of intimacy as longer-term residence, nor does it allow the observer to become integrated into the social round of the people under study. It does, however, give a longer-term view of a community, exposing patterns that take years to develop. It also makes it possible to obtain a different kind of feedback on one's conclusions. In the course of this study, I have published eight articles on various aspects of Jewish life in Copenhagen, including two in journals published by Jews in Denmark. All have been read by Danish Jews, and in some cases they have aroused some minor controversy. The comments I have received from readers in the community have materially improved my understanding of how Jews live in Denmark and how they perceive themselves.[4]

Wherever possible, I have tried to supplement this research with historical sources. Archival resources on the community tend to be limited, especially for the period before 1795. In the early twentieth century, however, a number of scholars began doing serious research on Jewish history in Denmark, and over the past few decades this work has accelerated sharply (see, for example, Arnheim 1950; Balslev 1932; Blüdnikow 1986; Blüdnikow 1991; Blum 1972; Borchsenius 1968; Feigenberg 1984; Margolinsky 1958a; Margolinsky and Meyer 1964). Some of this work has been done by outsiders, notably the studies of the rescue in 1943 (e.g., Elazar et al. 1984; Flender 1963; Kreth 1995; Sode-Madsen 1993; Yahil 1969). Most of it, however, has come either from Danish Jews or from Christians closely connected to Danish Jewry. The official Jewish Community publishes several serious journals of Jewish culture, each of which includes memoirs and historical articles. In addition, the Jewish Community's monthly newsmagazine, *Jødisk Orientering,* provides a record of Jewish events and activities going back almost a century. Together, these resources constitute an invaluable aid to an ethnographer, showing the origins and changing fortunes of contemporary institutions and customs. They also help in making sense of the often contradictory views expressed in interviews. By combining ethnographic with historical research, I have tried to develop as complete a picture as possible of Jewish life in Denmark and how it came to be.

This is not to say that I have gotten it all right. Any ethnographic study is necessarily a partial one, distorted by the limited perspective which any individual can gain on a community. While I have tried to speak with as broad a sample of Danish Jewry as possible, I have surely missed much, and the picture of the community presented here reflects those shortcomings.

PLAN OF THE BOOK

The structure of this book reflects the questions at the heart of understanding Jewish identity in late modern societies. Jewishness is a complex identity, incorporating such elements as a faith, a historical legacy, an ethnic community, a social network, an intellectual tradition, and a relationship with non-Jewish majorities. Each of these elements offers a variety of possible identities and worldviews with which individual Jews can engage. The meaning of each is also a subject of Jewish debate. In many places, therefore, including Denmark, the boundaries and nature of the Jewish world are impossible to define in any simple way.

Rather than beginning with a distinct model of the Danish Jewish community, therefore, this book will try to reach one, by considering the nature of Jewish experience through a number of different lenses. Chapter 1 sets the stage with a brief outline of Danish Jewish history. Then, in chapter 2, we examine Jewishness as a religious system, a set of rules, practices, and perspectives for relating the human and divine worlds. Jewish religious practice begins with a set of laws, laid out in the Bible and elaborated through millennia of talmudic criticism and debate. Like Jews everywhere, those in Denmark face the problem of integrating this tradition with their local cultural and social setting. What distinguishes this particular setting is the close engagement of Jews with the larger national culture, an engagement that creates dilemmas for living a Jewish life. How does one implement the taboo on consuming pigs, for example, when pork lies at the heart of the national cuisine? At a deeper level, how does one combine the ideas of divinity implicit in Jewish self-understanding with membership in one of the world's most aggressively secular societies? The Danish Jews have approached this dilemma through a policy of inclusion, by structuring ritual observance in a way that allows a variety of different approaches to Jewish experience to coexist. This chapter explores how that process works, discussing such practices as synagogue services, life-cycle rituals, and the observance of dietary laws.

Chapter 3 looks at Jewishness from a different perspective, that of social community. To what extent do Jews function as an effective group, forming and perpetuating social ties with one another? More to the point, how do they form subgroups, making social distinctions among themselves? Factionalism and opposition have characterized the Jewish world in Copenhagen since its inception, and they remain key elements of community life today. This chapter discusses some of the Jewish subgroups and their histories, focusing on what they imply about the nature of the overall community. In many ways, they attest to the community's vitality, its relevance to the widely varying experience of Jewishness implicit in a late modern social setting.

In chapter 4, we consider the Jewish community as a formal organization. To be Jewish anywhere involves more than an origin and a religious worldview; it implies an institutional connection, a relationship with an administrative body that oversees the manifold ritual and social activities associated with Jewish life. In Copenhagen, most of this activity falls under the authority of the Jewish Community of Denmark (*Det Mosaiske Troessamfund*), an institution formalized after emancipation as the official representative of Danish Jewry to the state. (In this book, I capitalize the term "Jewish Community," to distinguish this institution from the more general notion of the Jews as a group.) The Jewish Community funds and oversees the religious activities of the Great Synagogue, and it maintains a variety of educational and social service institutions. It also serves as a focus for community politics, as various subgroups contend for control over its administration. This chapter argues that Judaism's institutional base accounts for much of this community's persistent vitality. Places like schools, cemeteries, and nursing homes create a spatial dimension for Jewish identity, a physical setting in which to express individual notions of self and group.

Chapter 5 turns to the international dimension of Jewishness, the extent to which Danish Jews understand themselves in relation to Jews elsewhere in the world. Virtually all Danish Jews have relatives outside of Denmark, and most of them have traveled to Jewish communities in Israel, England, or the United States. As a result, these distant communities exert a powerful influence on Jewish identity and religious practice in Copenhagen. This chapter discusses some of the mechanisms through which Danish Jews interact with the larger Jewish world, including Zionist associations, travel, educational programs, refugee assistance, and family contacts. These contacts have important effects on Danish Jewry, some of them apparently contradictory. While romantic images of Israel, for example, have intensified Jewish identity among many youths in Denmark, they have also undermined the authority of local notions of Jewish authenticity.

After looking at Jewishness as an international identity, we then turn to its national context. Chapter 6 considers how Jewishness is conceived in Danish culture and how that conception shapes Jewish experience and self-understanding. Most Jews in Denmark are, after all, also Danes; Danish values and attitudes are as much a part of their experience as Jewish ones. Yet most Danes know almost nothing about Jews, and what they do know is largely limited to a few historical and cultural stereotypes. This situation presents a problem for Danish Jews, who must understand themselves simultaneously in two very different cultural frameworks. They must also face the difficulty of being different in a society that strongly values homogeneity. By any reasonable measure, the level of anti-Semitism in Denmark is extraordinarily low; at the same time, however, Danish culture generally frowns on difference, strongly stigmatizing individuals who mark themselves off as special or unusual. For many Jews, these patterns create a dilemma. As Jews, they see themselves and are seen by others as different, while as Danes they want to be just like everyone else. The difficulty of combining these identities poses real problems for many Jews, and it helps produce a wide variation in the ways that individuals construct their own notions of what it means to be Jewish.

Chapter 7 explores one final dimension of Jewishness in Denmark, its relationship with the past. To be Jewish is in part to be a product of history, a person who shares for good and bad in a long historical legacy. An essential part of constructing a Jewish identity involves formulating a personal understanding of the meaning of that legacy. This chapter looks at how Jews do that in Denmark, focusing particularly on the construction of one event: the dramatic rescue of the Jews from the Nazis in 1943. This event has become an emblem both of Danish resistance during the occupation and of the incorporation of the Jews into the national body. As it has been recounted over and over, both orally and in journalistic accounts, the story of the rescue has assumed a fairly standardized form, one that presents Danes as selfless heroes, Germans as demonic villains, and Jews as passive cargo. On the ground, of course, the events and actors were much more complicated, and individual experiences of the rescue varied greatly. This chapter looks in detail at four narratives of the rescue, told by four very different members of the community. The differences in these narratives illustrate some of the differences in the Jewish experience in Denmark, and the variety of ways in which individuals negotiate the dilemmas of modern Jewish identity.

In chapter 8, we conclude with a discussion of what all this means for our understanding of community in late modern societies. As noted above, conceptions of community in the social sciences have often imagined it in terms of *gemeinschaft,* seeing communities as unified and bounded groups whose members share a common worldview. Early anthropologists projected such an image onto the isolated tribes where they conducted their fieldwork, and later ones projected it onto peasants and urban villagers. This notion, however, was fundamentally incompatible with modern social processes, which tend to break up bounded groups and local identities. Many social scientists therefore predicted that modernization would see the ultimate extinction of community in human life. In recent years, however, two trends have called this view into doubt. One has been the growing realization that even the sites of classic anthropological fieldwork were never as bounded or unified as Westerners thought they were. The other is that community, rather than disappearing from the modern lexicon, has become increasingly omnipresent in public discourse. Community, in other words, didn't exist where we thought it did, and refuses to die where we thought it would. The same sort of realization has dogged anthropological conceptions of culture and society, which have come under attack from postmodern theorists.

This chapter suggests that Anthony Cohen's notion of the community as a symbolic field allows a resolution of some of these problems. Cohen emphasizes the extent to which members of a group connote their symbols in widely differing ways; what unites them is not so much shared understandings as a common commitment to this particular body of symbolic referents. The chapter applies this view of community to the Copenhagen Jews and suggests implication for the futures of this and other western ethnic communities—futures that may be far less bleak than social science has often suggested.

Author's Background

One of the defining features of recent theory in anthropology has been its interest in the interaction between the experiential worlds of anthropologists and the images of culture that they create (cf. Geertz 1988; Kuper 1988). Anthropologists are not scientific instruments, recording objectively perceptible cultures and replaying them dispassionately for their readers; they are human beings who bring individual personalities and motives both to the fieldwork encounter and to the writing of ethnography. This observation applies to this study as much as any other. While I have tried to present as accurate a picture as possible of Jewish life in Copenhagen, I have done so from a distinct perspective, and another observer might paint a significantly different one. A few words on my own background, therefore, may help readers evaluate some of the information in the pages to follow.

My religious background is in some ways well matched to a study of Danish Jewry. My mother's family emigrated from western Denmark to the United States in the late nineteenth century; my father's parents came to New York from a Polish shtetl in the early twentieth. As a result, while I was raised as a Unitarian, both Danish Lutherans and Eastern European Jews were part of my family experience from an early age. At a somewhat later age, I had the good fortune to marry a Jewish woman. Today I am a member of both a Reform temple and a Unitarian church, and my family attends holiday services in each. In this sense, the questions raised in this study about the interaction of Danish and Jewish identities have an echo in my own experience.

Some scholars have suggested that being Jewish oneself is close to a necessary precondition for doing fieldwork among Jews, if only because a knowledge of Hebrew and Jewish ritual practice is so hard to come by any other way. Non-Jews also face difficulties in being accepted by their field subjects that Jews do not. While I see the logic in this position, I disagree with it, and I have not treated my religion as a disability in my fieldwork.[5] When the subject came up in interviews, I told people that I was not myself Jewish, although I had connections with Judaism through my father and my wife. This did not make my fieldwork easier; even Jewish immigrants tend to feel excluded from this community, and as a nosy non-Jew I often felt a disheartening cold shoulder during my participant observation. At the same time, however, if my knowledge of Hebrew is limited, and if my understanding of Jewish theology is unsophisticated, the same could certainly be said for the great majority of the Danish Jews. And if I sometimes felt snubbed and isolated by the Jewish Community, so too did very many of those whom I interviewed. Like most anthropologists, I felt my outsider status as both a blessing and a curse, something that locked me out of some areas of the Jewish world but heightened my sensitivity to others.

The course of this fieldwork has taught me a great deal about the Jewish experience, and I am grateful for what I have learned about my own religious heritage. That said, I should note that this background is not the reason I became interested in the Copenhagen Jews. I neither planned this study nor experienced it as a project of personal discovery. The reason actually had to do with an elderly cousin of mine, Gerda Jeppesen, who runs a farm on a lovely flyspeck of an island called Fur in northern Denmark, and whom my wife and I became very close to during our first fieldwork on the nearby island of Mors. Gerda was one of the first Danes to find out that my wife was Jewish, and she made a point of relaying to us any Jewish item that found its way into the local papers. We visited Gerda quite frequently in 1990 and 1991, and many times she greeted us with a handful of clippings about the Danish Jews. Eventually she wore me down; when I finished my study of Danish Lutheran sects in 1995, it seemed fairly natural to look into the Jewish community. I actually knew almost nothing about them when I began the study, beyond what I had learned from Gerda. Had I known more—had I realized how different this community would be from others I had studied, and how many problems of fieldwork and interpretation it would present—I would very likely have done something different. Ignorance can be a blessing at times.

One other point about my perspective in this book. Much of the social scientific work on Jews has come from a frankly activist viewpoint; authors have sought not merely to understand Jewish communities, but also to find solutions to the problems they face. They have focused particularly on the "crisis of assimilation," the extent to which the survival of the Jews as a people is threatened by their engagement with host cultures in the West. While I have the deepest respect for this type of research, I have not attempted to do it here. I discuss a number of features of the Danish Jewish community that many of its members regard as dangers. I have not, however, organized the text around these issues, nor have I made any suggestions for ameliorating them. Partly this is because I think the community is in much less danger than it might appear. Mostly, however, it is because I feel that it would be presumptuous. The job of an anthropologist is to try to understand the cultural world of the people he encounters; the right to change it is theirs. I have neither the expertise nor the moral standing to say what is wrong with this community or how it should be fixed. Fortunately, the Danish Jews have no shortage of opinions on this subject, and they have the power to change things if they wish to. I can at best provide an outsider's perspective on the community, which they can note or disregard as they wish.

A Note on Pronunciation and Usage

The Danish language includes several letters not used in English, and it uses different pronunciations for some of those the two hold in common. This can make Danish words look rather impenetrable to English speakers. The table below is intended to help English speakers pronounce some of the Danish terms they come across in the text. The pronunciations are only approximate, as Danish phonology is different from English, and pronunciation varies from one region of the country to another.

Danish character	Pronunciation
Æ, æ	Halfway between a long *a* and a long *e*; like the *ea* in *bear*
Å, å	Close to a long *o* in English
Aa, aa	Same as å; the letter å is a modern replacement for the traditional double *a*. Thus *paa* is the same word as *på*, and both are pronounced much like *Poe*.
Ø, ø	Like the *u* in *purr*
Y, y	Similar to a long *u* in English. The Danish *y* is never used as a consonant.
J, j	Like the English consonant *y*

Danish nouns and adjectives can also be confusing for English readers. The ending of an adjective, for example, reflects the gender and number of the noun modified, as well as whether a definite or indefinite article precedes it. This can make spellings appear inconsistent; "Jewish Community" would be *Mosaisk Troessamfund*, whereas "The Jewish Community" would be *Det Mosaiske Troessamfund*. Confusion can also result from the Danish pattern of using a suffix to indicate the definite article if there is no adjective; thus, "a Jew" is *en jøde*, "the Jew" is *jøden*, and "the tall Jew" is *den høje jøde*. As the last example indicates, Danish also uses different rules for capitalization, so that many words capitalized in English are lower-case in Danish.

In order to maintain the confidentiality of informants, I have used pseudonyms and altered details of appearance and life histories when discussing individuals. The exception to this rule is for public figures, such as rabbis and officers of the Jewish Community, whose identities are common knowledge and would in any case be impossible to conceal. When discussing such figures in their official capacity, I have referred to them by name.

The Community
in Time

Wulf Wallich was a child of the sixties, and he gave the rabbi fits. Like a number of young, progressive Jews in the eighties, he thought that the Jewish community was behind the times. Its leaders seemed to think they were in a seventeenth-century shtetl, not modern urban Copenhagen, and they didn't want to adapt to the changing world around them. He and his friends made a lot of trouble for the heads of the community, even taking them to court on accusations of skimming funds. But what really irritated them was his hair—straight out of the sixties, with braids and ribbons and all, not what they considered appropriate for a respectable Jewish businessman. It was embarrassing enough during the week, but the rabbi, Gedalia Levin, was insulted that Wallich and his friends came to the synagogue that way on Saturdays. He finally published a rule forbidding such hairstyles in the synagogue, virtually throwing down a gauntlet to troublemakers like Wallich. And inevitably, one Saturday morning in the late eighties, Wallich took it up.

The occasion was the anniversary of his mother's death, her yahrzeit. The occasion called for a mitzvah, and Wallich expected one; according to custom, he should have been called to the central bimah during the service, and the cantor should have sung a number of verses in his name. Rabbi Levin, however, took one look at Wallich's dramatic hairstyle and decided otherwise. He refused to call Wallich to the bimah, even when Wallich pointedly asked him to. The cantor, following the rabbi's instructions, refused Wallich's demands to sing for him. Enraged, Wallich finally stormed to the bimah and began reading from the Torah himself. The synagogue burst into disorder as a furious debate broke out immediately. Some called Wallich a hero, a soldier against the archaic rigidity of a domineering elite; others denounced him as the epitome of a self-centered generation that wanted to squander centuries of tradition. Tempers rose so high that the dispute ended in court, touching off a lengthy government investigation into reforms and financial scandals in the Jewish community. For decades afterward, both supporters and opponents of congregational reform cited Wallich's behavior as evidence for their cause.

The controversy over Wallich's hair illustrates a central question about the history of Danish Jewry. The events themselves sound familiar to any observer of Western religion in the second half of the twentieth century; the aesthetic manifesto of the counterculture led to decades of confrontation over styles of dress and worship, usually reflecting deeper social and ideological oppositions within congregations. Recent decades in Denmark have seen a number of similar clashes, though seldom quite so dramatic. The names of the principals, however, sound unfamiliar, even archaic. Indeed, they should—Gedalia Levin died in 1793, when Christian VII was king of Denmark and George Washington president of the United States. The hairstyle in question was not dreadlocks or cornrows; it involved a long pigtail with colored ribbons and a comb, a style that had begun to replace wigs in Denmark around 1760. The battle in the synagogue came to be known as the Wig War, and it received both serious and satirical newspaper coverage in the late 1780s (see Feigenberg 1984: 75–77; Margolinsky 1958a: 50–51).

What do such parallels tell us about the nature of history among the Danish Jews? Is it merely the recurrence, over and over, of a certain set of characteristic oppositions, a retelling under different circumstances of an enduring story about the essence of Jewish culture? Writers have sometimes portrayed Jewish history this way, suggesting that some structural properties of the Jewish experience—the Diaspora condition, the talmudic tradition, ongoing anti-Semitism, and others—transcend the specific times and locations within which particular Jews have lived. Other writers have argued otherwise, noting the tremendous differences in the conditions under which these apparently similar events take place. When Wallich took his case to court, for example, he did so as a non-citizen, and he called on the government to investigate a Jewish community that existed largely outside of Danish law. These circumstances made his actions, and their consequences, qualitatively different than disputes over dress and hairstyle in the 1960s. How much do each of these perspectives tell us about the relationship between the past and the present in contemporary Jewish Copenhagen? Where is today's Jewish culture an extension of its history, and where does it represent something genuinely new?

In this chapter, we will sketch the history of the Copenhagen Jews from the seventeenth until the late twentieth century. We will consider the kinds of community structures, social oppositions, and cultural trends that have characterized the group over that time. As we do so, we will find both continuity and change, recurring patterns and evolving circumstances, which make today's Jews both extensions of and departures from their common past.[1]

Early Settlement, 1673—1784[2]

The Copenhagen Jewish community dates back to the 1670s, when a handful of Jewish merchants and tobacconists began moving to the city from northern Germany. Jewish immigration to Danish territory had begun back in 1628, when King Christian IV had attempted to establish a new commercial center in the town of Glückstadt in Slesvig. To attract capital, he had offered religious privileges to Jews, and a number had moved there from Hamburg and Altona. The new center never took off, though, and within a few years all of the Jews had left again. More permanent immigration began only after Denmark's disastrous war with Sweden in 1558–1560. The war left the country financially desperate, and the crown looked to immigration as a means of bringing in new capital (Blüdnikow and Jørgensen 1984: 23–26). Beginning in 1673, therefore, it began extending residence permits to wealthy Jews who wished to settle within the kingdom proper. The few who did so scattered themselves around the country, creating small Jewish circles in Fredericia, Nakskov, Århus, Køge, and Ribe. Eleven households settled in Copenhagen, giving the city a Jewish population of nineteen by 1682.[3]

The permits allowed only residence, not religious concessions; Denmark's state Lutheran church forbade all competing religions, and even foreign ambassadors had sharp restrictions on their religious practice.[4] Jews were not forced to convert, but neither were they allowed to hold any sort of religious observances of their own. Burial in Christian cemeteries was likewise forbidden, and Jews who died in Denmark were transported to Altona for their funerals (Margolinsky 1958b: 42). As the communities became established, however, and as their economic services became more valuable, they slowly gained some limited religious privileges. In Copenhagen, Jewish observance first became legal in 1684, when a court jeweler named Meyer Goldschmidt received permission to hold religious services in his home. The conditions for observance were quite strict—the services had to be held behind closed doors, and no sermon was to be given—but Jews from around the city were allowed to attend, an unusual concession for any nonconformist religion at the time (Balslev 1932: 14–15). In 1694, the Jews received permission to establish a cemetery just outside the northern boundary of the city. These concessions made possible a functional, if limited, Jewish community, and the Jewish population grew slowly but steadily over the next half century—from eleven households in 1682 to 65 in 1726, and to 120 by 1754.[5]

Even at this early date, important divisions existed among the Copenhagen Jews; indeed, these differences and the legal conditions attached to them predated Jewish immigration into the city. Christian IV had directed his patronage specifically toward Sephardic Jewry, descendants of families who had left Spain during the expulsions of 1492. After settling for a time in Portugal, many Sephardim had relocated to the Netherlands, and some had become important financiers in northern Europe. It was these Jews, known as "the Portuguese," whom Christian invited to Glückstadt in 1628 (Balslev 1932: 5). His letter of privilege was addressed to "the Portuguese Nation," and it recognized this group's opposition to the Ashkenazi Jews of Germany. As long as the Portuguese behaved as good merchants, it said, German Jews would be barred from the city. This legal preference remained in force for more than a century, as royal policy repeatedly granted the Portuguese rights that it did not extend to the Germans.

In Copenhagen, on the other hand, German Jews held the upper hand in practice right from the start. Whereas the crown had hoped to build a commercial center in Glückstadt, its plans for the capital turned on industry, particularly the manufacture of tobacco products. The first Jewish immigrants were cigar makers from Hamburg, while subsequent settlers included German-Jewish jewelers and other small-scale manufacturers. A small Portuguese contingent did live in the city, but relations between the two groups were distant and strained. Government policy never granted the Germans privileges as a distinct community, the way it had earlier with the Portuguese; since the individual privileges it granted tended to go to Germans, however, they came to dominate Jewish life in Copenhagen. When Meyer Goldschmidt began holding services in 1684, for example, he invited only German Jews. The Portuguese, who used a somewhat different ritual, had no legal services until 1695 (Balslev 1932: 15). The cemetery likewise largely excluded the Sephardim (Margolinsky 1958b: 42). The distinctions between the two groups retained legal force until 1814, and their social significance continued into the twentieth century.

An even more significant distinction turned on the matter of wealth. The crown admitted Jews to Denmark for their economic value, and the residence permits went only to Jews of demonstrated fortune or commercial prospects. Without the permits, Jews had no right to live in the kingdom; they were routinely expelled, fined, or jailed if discovered by authorities. The government nourished a particular enmity for peddlers, itinerant traders whom it saw as a threat to rural trade and social order. Throughout the seventeenth and eighteenth centuries, however, significant numbers of poor Jews entered the country illegally, and many of them settled in Copenhagen. These Jews lived in the city's slums under difficult conditions, disliked and often persecuted by non-Jews and unable to turn to the authorities for protection. Their relations with the established community were mixed. The wealthier Jews resented the unfavorable attention that the peddlers attracted to their faith; at the same time, the poorer Jews were coreligionists and often relatives, and they constituted a valuable pool of workers and potential spouses. Accordingly, the legal Jews tended to act as a patron class, often interceding legally or financially on the poorer Jews' behalf. Even centuries later, after legal residence had been long established for all Jews in Denmark, some of these wealthy families retained an association of aristocracy within the Jewish community.

Factionalism also developed along religious lines as small differences in observance formed the basis for cliques and antagonisms. In Goldschmidt's prayer room, for example, services followed a Polish format; this custom, or *minhag,* was standard for Goldschmidt's home city of Hamburg. In the early 1700s, Meyer Levin, a Jew from the city of Emden, objected to this style and began holding services in his own home according to the Emden *minhag.* Those who attended became a sort of opposition to Goldschmidt's dominance in the community, and a series of increasingly scandalous clashes between the two groups developed. The Levin circle hired its own religious leader, Marcus David, who became a competitor of the established rabbi, Israel Ber. When Levin's followers stopped paying their dues to the community, they were officially expelled, and Ber denied them burial rights in the Jewish cemetery. The conflict continued until 1728, when a massive fire destroyed both Goldschmidt's and Meyer's houses, and Ber left the city for Fredericia (Margolinsky 1958b: 43). Such divisions arose constantly in the Copenhagen community, which usually had a number of different and opposing prayer rooms operating at any given time.[6]

Rich or poor, Ashkenazic or Sephardic, Hamburger or Emdener, none of these Jews was Danish, either in his or her own eyes or those of the larger society. They formed a distinct group, with a distinct culture, language, and legal status. All of the German Jews spoke Yiddish, and few appear to have spoken more than a smattering of Danish; their interactions with Christian neighbors were restricted mainly to business activities. The Danish guilds barred them from membership, closing off most occupations outside of commerce, and Jewish doctors could not treat Christian patients. The Danish legal system imposed a variety of special requirements on Jews, many of them based on its antagonism to the Jewish faith. A law of 1725, for example, forbade Jews to employ Christian servants, for fear that the servants might be either seduced or converted by their employers. Likewise, Jewish men were liable to imprisonment and subsequent banishment if they impregnated a Christian woman. Perhaps the most resented requirement concerned legal oaths. A Jew swearing an oath in a court case had to do so in the synagogue, dressed in ceremonial robes with a rabbi and other functionaries present; the oath itself involved a lengthy set of pronouncements designed to counter the Jew's presumed disposition to lie (Balslev 1932: 18–20; Blum 1972: 23). Popular attitudes toward Jews echoed this sentiment. Anti-Semitism in Denmark was significantly milder than elsewhere in Europe, and Jews were seldom visited with physical assaults. They endured constant scorn and suspicion from their neighbors, however, and street toughs frequently cast insults or stones at them. The Church denounced them as enemies of the faith, while the guilds accused them of smuggling in foreign goods. And while never actually accused of ritual murder, a charge that had occasioned persecution many other places in Europe, the Danish Jews seem to have been associated with it in the public eye. In 1699, an impoverished Christian woman visited Meyer Goldschmidt and offered him a baby for sale, to be used as a blood sacrifice in the making of Passover matzohs (Bamberger 1983: 22–23). Even Goldschmidt, one of the wealthiest and most prominent of the Jews, and a personal favorite of King Christian V, remained in common eyes a potential murderer of children.

This distinctive status, for all its difficulties, did confer a degree of autonomy on the Jews as a group. A council of elders administered the community according to Mosaic law, and their decisions were largely respected by the Danish authorities. If individual Jews defied their authority, the elders could impose a punishment of excommunication. The council also probated and administered Jewish wills and estates, and the community maintained its own poor-relief programs. Marriages were regulated by Jewish religious law, while Jewish burial societies arranged for the interment of the dead. Despite its general distrust of Jews—and its occasional efforts to convert them to Christianity—the Danish government regarded the Jewish community as a distinct and legitimate entity, one that had the authority to manage its own internal affairs.

Though they stood outside of Danish society, the Jews of Denmark did live in its midst, and as time went by they gradually appropriated elements of Danish culture. While early Jews dressed and groomed distinctively, for example, with caftans, long beards, and round fur hats, they soon began following Danish styles. In 1710, the use of wigs for men and hair ribbons for women had become so common that the community instituted a tax to try to discourage it. Danish Jews developed a distinctive dialect of Yiddish that incorporated Danish words and references. Over time, particularly among the well-to-do, Danish standards of daily life and ritual practice became the standards against which the propriety of Jewish life was measured. By the late eighteenth century, the noisy ragtag processions that customarily journeyed to the cemetery for Jewish funerals became a matter of outrage among many leading Jews (Margolinsky 1958b). Likewise, the Jews' messy and ramshackle houses became an increasing embarrassment in a land that prized tidy homes. This movement toward Danish culture had distinct limits, and both intermarriage and conversion to Christianity were extremely rare before the nineteenth century. But for all its exclusion and autonomy, a century and a half of residence in Copenhagen gradually transformed the Jewish community. By the late eighteenth century, one could speak of its members not simply as Jews living in Denmark, but in a real sense as Danish Jews.

THE REFORM ERA, 1784—1814[7]

From their first arrival in Denmark until the last part of the eighteenth century, Jews maintained a consistent status as an alien minority in a Christian nation. Governed by their own institutions, restricted to their own social circles, separated culturally from the world around them, they were sometimes referred to as a "state within a state." Their legal status derived not from any inherent rights or principles but from a confused and inconsistent series of royal decrees and permissions. Between 1784 and 1814, however, this condition changed radically. Through a series of legal and religious reforms, the Jews became regular citizens of the Danish state. In doing so, they gained many of the privileges of citizenship, and at the same time they lost the cultural and legal autonomy that had defined the community since the 1600s.

The impetus for this process of reform came from two directions. On one side, the Danish state entered a new period in 1784, when the progressive Crown Prince Frederik (later Frederik VI) took control of the state. The sixteen-year-old Prince Regent began a thorough reform of Denmark's system of law and land tenure, aiming to replace a largely medieval legal system with a rational enlightenment model (Gold 1975). This project made the Jews an object of special concern, in part because their special status clashed with the legal standardization implicit in the reforms. In addition, the Jews' exclusion from various professions provided an opportunity for Frederik to attack competing power structures within the kingdom. The guilds, for example, which determined who could enter skilled trades in Denmark, presented a barrier to Frederik's plans for liberalizing trade; the question of admitting Jews offered an opening through which Frederik could break their monopolies. Combined with what appears to have been a sentimental interest on Frederik's part in the welfare of the Jews, these issues led the government to thoroughly reexamine the position of the Jews in Danish society.

At the same time, trends in the Jewish world augured for a greater connection with the larger culture. Denmark's neutrality in the wars of the late eighteenth century had created abundant opportunities for merchants, leading hundreds of German Jewish merchants to move to Copenhagen. The Jewish population of the city grew from about 700 in 1760 to 1,200 in 1784, and it would reach 1,500 by 1800. As the community grew, its medieval administrative system became increasingly unequal to its task; disputes arose about finances and practices, and dissatisfied members looked to the government for help. They also looked to Germany, where the Jewish Enlightenment movement led by Moses Mendelssohn was calling for the incorporation of Jews into the states in which they lived. In the second half of the century, therefore, a small but very active group within the Jewish community began pushing to modernize the Jews' relationship with the Danish state. This group—which included Mendelssohn's brother in law, Moses Fürst—never constituted a majority in the community, and indeed its recommendations met strong negative reactions whenever they became public. But by working together with a reform-minded government, it managed to transform the community radically over the course of two generations.

The earliest and least controversial reforms involved the skilled trades and professions. The possibility of admitting Jews to the University of Copenhagen had been debated for decades, since a Jew named S. J. Wolff had fulfilled the requirements for a doctorate in medicine in 1761. The university had refused to award the degree, citing a part of its constitution that forbade giving degrees to those with "false knowledge." In 1788, the offending paragraph in the constitution was removed, making the university and the professions officially open to Jews (Balslev 1932: 35–37). Concern over the skilled trades came later, but it had a much larger impact. In 1788, to the outrage of the guilds, Frederik issued a decree requiring masters to accept Jewish apprentices. The decree imposed stiff fines for refusal, and the crown enforced them stringently. Acceptance of the offer was limited at first, since it imposed a hardship on Jewish families—masters could not provide kosher food, which meant that Jewish apprentices had to be fed by their families. In 1793, however, a joint group of Christian and Jewish philanthropists established a foundation to fund board for Jewish apprentices, and the effects were impressive. Within forty years, a community that had depended almost entirely on commerce for its income had a quarter of its men working in the skilled trades. These reforms effectively released Jews from their occupational ghetto, leading to their first real integration into the larger Danish society.

Within the confines of the Jewish community, meanwhile, reformers began to challenge the traditional authority of the rabbi and elders. The Wig War, for example, led to a civil suit in the Danish courts, in which a number of features of community administration came under scrutiny. Reformers accused the elders of incompetence and corruption in their handling of community finances; they focused particularly on the probating of wills and the care of orphans, which they said the elders had milked for financial gain. They also objected to mandatory community obligations, such as requirements to take in and feed foreign Jews visiting the city. Reformers regarded such policies as unfair and demanded that the crown put an end to them. Their success in court was mixed, as the crown recognized the elders' right to impose their authority on internal matters. At the same time, though, the trials showed a number of irregularities in the community's finances, and the persistence of the reformers forced the community leaders to compromise. In 1791, the two sides established a commission to oversee community bookkeeping; they also agreed to make feeding of foreigners voluntary, and they found a diplomatic solution to the wig problem. The compromise took some of the independence and authority away from the Jewish administration, but it quieted the reformers and left the essence of the elders' authority intact.

These lawsuits had cracked the doorway to congregational restructuring, however, and the government lost no time in trying to push it all the way open (Gold 1975). In 1795, the crown established a commission to investigate the status of Jews, composed of government ministers and leaders of the Jewish reform movement. Their report, published the following year, recommended a thorough restructuring of the status of Jews in Denmark. It began with a preamble that decried the existing Jewish legal system as confusing and obsolete. Women had inferior rights to those of Danish women, and laws about marriage and inheritance followed archaic forms. Moreover, said the report, the Jewish "state within a state" reinforced cultural divisions and prejudices that made Christians and Jews regard each other with suspicion. The solution lay in regularizing the status of Jews, subjecting them to the same laws as their neighbors except in the religious sphere. Consequently, the report proposed a detailed set of administrative reforms, which would have a transforming effect on the Jewish community. It spelled out new systems for electing officers, choosing rabbis, keeping records, conducting funerals, regulating marriage, and probating wills—all oriented toward conforming Jewish practice to the principles of rational administration then popular among Danish ministers.

The report, published in August 1796, set off an unprecedented uproar in the Jewish community. The vast majority of the Jews regarded the proposals as a wholesale assault on the integrity of their community, one that would destroy Jewish cultural distinctiveness as well as administrative autonomy. In February 1797, the community's leaders presented a petition of protest to the crown, signed by 167 heads of Jewish households. In the face of such opposition, the crown shelved the proposals, which would ordinarily have become law upon publication. It did not, however, forget them. Over the course of the next twenty years, a variety of reforms occurred in the Jewish community, most of them prefigured in the 1796 report. By 1814, when a royal proclamation officially declared the Jews to be Danish citizens, the structure and orientation of the Copenhagen community had changed profoundly, its emphasis on the maintenance of traditional forms replaced with an concern for integration and modern rationality.

Many of these changes bore the signature of Mendel Levin Nathanson, a businessman and economist who would later become well known as the longtime editor of the newspaper *Berlingske Tidende*. As the leader of the community board, and with the support of his influential grandfather, David Anselm Meyer, Nathanson initiated a flurry of reforms between 1805 and 1814. Some of these reforms sought to even out class inequities in the community. In 1806, for example, Nathanson abolished the tax on kosher meat that had provided much of the administration's funding, replacing it with a more progressive tax on income. He also led the establishment of two schools for poor Jewish children, a boys' school in 1805 and a girls' school in 1810. Other reforms harmonized Jewish practice with prevailing Danish customs. In 1810, the community began keeping the same kinds of vital statistics records, called *ministerialbøger*, as Danish churches. In 1811, Nathanson changed the community's burial customs to bring them more in line with Danish practice. The often ragtag funeral processions were made more seemly, various undignified activities at the cemetery were abolished, and most importantly, the length of time between death and burial was extended from hours to days. In 1812, Nathanson commissioned a new, Reform-oriented instruction book in Jewish religion, to serve as an analog to the catechisms used in Lutheran confirmation classes. These changes often aroused anger in the congregation, most of whose influential members remained staunchly conservative; even Nathanson's grandfather objected vigorously to the burial reforms. Their effects, however, were irresistible, as despite itself the community turned its face more and more toward the larger society.

Nathanson's ability to make such headway derived to some degree from royal patronage; the crown favored his views heavily and installed a sympathetic community board to support him in 1809. Crown Princess Caroline herself became the patroness of the girls' school, which was named in her honor. The success of the reform movement, however, reflected a general community fragmentation that made organized opposition very difficult. The main synagogue had burned down in 1795, and in the years afterward religious worship was scattered among a number of private homes. This dispersal augmented the factionalism already present in the community, making large cooperative projects almost impossible. In 1799, for example, the community purchased land in Krystalgade for a new synagogue; before construction could begin, however, a dispute over who would have rights to the choice seats became so intense that the entire project was put off indefinitely. The feuding groups could come together at moments of crisis, as they did in 1796, but on an everyday level their disagreements predominated. As a result, the incremental reforms introduced by Nathanson and his supporters met little organized opposition. While the vast majority of Copenhagen Jews unquestionably opposed any major changes, Nathanson's small and committed group managed fundamentally to alter the conditions of Jewish life between 1805 and 1814.

Reluctant as many Jews were to join the larger society, Denmark's Christians were no more enthusiastic about letting them in. For all of Frederik's interest in the Jews, most Danes continued to regard them as a suspicious and undesirable outgroup. These feelings intensified after a series of national catastrophes in the first decade of the new century. Denmark's ill-fated alliance with Napoleon led to the bombardment of Copenhagen by Britain in 1807 and subsequently to the loss of the Danish navy and much of its merchant fleet. With commerce and credit cut off, the economy fell to pieces, and some of the resulting anger was directed at the Jews. This sentiment culminated in 1813 with the "literary Jewish feud," (*litterære jødefejde*), a spasm of attacks on Jews in the popular press that seized the public imagination between May and August (Borchsenius 1968: 53–58). The feud began with the translation into Danish of an anti-Semitic tract by Friederich Buchholtz entitled "Moses and Jesus." A flurry of articles and pamphlets followed, some supporting the book and others defending the Jews. The attacks moved quickly from general complaints about Jews to specific accusations against the Copenhagen community, and they came to focus on the charge that Jewish financiers had caused the credit crisis by moving money out of the country. This charge met with detailed refutations from Nathanson and other Jewish leaders, as well as from a number of prominent Christians. After a few months the storm passed over; by the end of the year, the topic had disappeared from the press and popular discussion. Its intensity, however, spawning at least twenty pamphlets and five short-lived newspapers in its brief span, testified to the enduring animosity toward Jews in parts of the Danish population.

The king made no public statements on the feud, which largely burned itself out by the end of August. In its wake, however, he expressed his position as only an absolute monarch could. Frederik had ordered the chancery to draft a decree proclaiming Jewish citizenship back in 1806, and a preliminary version had been completed in 1807. The national crisis had put the issue on hold since then; as the literary battle raged, however, Frederik ordered the proposal revived. The document appeared without dramatic fanfare or a lofty title—it continues to be known only as "The Decree of March 29, 1814"—but it introduced the most profound transformation of the Copenhagen Jewish community since the seventeenth century.

The Decree of March 29, 1814[8]

In histories of the Danish Jews, March 29, 1814, stands out as an epochal date. Authors describe it as the date on which Jews became Danish citizens, the date when full civil rights were extended to the Jews, the date when the generally positive attitude of the Danish crown toward the Jews became a matter of statute rather than inclination. The community still celebrates anniversaries of the date, and for the one hundredth and one hundred fiftieth anniversaries the Mosaisk Troessamfund published major commemorative volumes. There is little doubt that Jews of the time saw it in similar terms, and that many responded to it with jubilation.

Given this response, it can be startling to read the decree itself, and to observe how little of the actual text concerns rights for Jews (Gold 1975). Indeed, the only explicit right granted concerns employment: in the first paragraph, the decree orders that "Those of the followers of the Mosaic religion, who are born into Our kingdom of Denmark, or have received permission to settle therein, shall be permitted to enjoy the same rights as Our other subjects to support themselves in every lawful way." The rest of the document involves a systematic repeal of a variety of special privileges and powers held by Jews and the Jewish Community, as well as the imposition of a number of administrative rules and requirements for the conduct of Jewish worship. The main thrust of the decree, both in tone and weight of words, is the dismantling of the Jewish Community as a secular administrative agency.

The decree consists of twenty paragraphs, which lay down three basic kinds of provision. One concerns the right, mentioned above, to engage in any lawful occupation without hindrance based on religion. To a large degree, this right already existed; Jews had been admitted to the guilds and professions decades earlier. The provisions of the decree, moreover, would prove tricky to enforce, as Danish employers found numerous subtle means of denying jobs to Jews when they wanted to. It would certainly be wrong to understate the importance of the decree, which for the first time ascribed positive rights to Jews based on their status as residents of Denmark. This importance was, however, largely a matter of symbolism; its immediate effects on Jewish social life, if not negligible, were practically quite limited.

The other two types of provision, by contrast, produced quick and perceptible effects. One set focused on the elimination of the "state within a state," the special administrative procedures that had characterized the Jewish community since its inception. All administrative and business records, for example, were henceforth required to be kept in Danish, using Gothic script and the Gregorian calendar. Marriage documents, wills, and other contracts had to follow Danish legal formats, or they would be considered invalid; likewise, marriages, divorces, and estates were to be administered according to Danish law, not Jewish law. (This meant, for example, that the Community had to recognize interfaith marriages approved by the Danish government, and that rabbis could no longer issue divorces.) The decree revoked the power of excommunication, and it ordered specifically that "No representative, priest, or any member of the Mosaic religious community may dare to obstruct or disturb any follower of the Mosaic religion in the performance of any lawful trade . . . or assume any power over their domestic condition." Taken together, these provisions effectively destroyed the secular power of the Jewish Community, arrogating its most important civil functions to the state and forbidding it to interfere in nonreligious matters. Moreover, it eliminated many of the cultural mechanisms through which Jews had maintained their distance from Danish culture. To run their businesses, Jews now had to learn Danish language, administrative practice, and time reckoning. To register births, deaths, and other life transitions, they had to engage with the bureaucracy of the Danish state. Before 1814, most Jews had lived with very little contact with the world outside their community; after 1814, that world was a basic part of their existence.[9]

The other type of provision addressed the format and organization of Jewish religious worship. As with civil matters, these regulations tended to impose state power on Jewish religion and to reshape it in forms more like the state Lutheran church. The decree ordered that all Jewish clergy in Denmark—described as priests, not rabbis—should be authorized by the state, with one chief priest to live in Copenhagen.[10] The crown would determine clerical salaries, and the performance of "priestly acts" by unapproved individuals was strictly prohibited. The decree also prohibited the establishment of synagogues without the specific permission of the state. Perhaps most importantly, the decree ordered a new system of religious instruction and confirmation for Jewish boys and girls. It required that all children be instructed using the Reform-oriented catechism developed by Nathanson, and that at age fourteen they be tested to confirm that they understood its contents. Following the test, the children were to take vows promising to abide by the rules laid out in the text. These regulations represented something unprecedented in the history of the Danish Jews—a direct intrusion by the state into the conduct of Jewish worship. Earlier regulations regarding the Jews had focused mainly on immigration and occupational issues; where they touched on religious observance, they had generally involved limits on where and when it could take place. In the 1814 decree, however, the state dictated the institution and structure of specific rituals, and it approved a specific formulation of the Jewish faith.

The Decree of 1814 represented, then, a fundamental turning point in the history of the Danish Jews. After its publication, the isolated world in which most of Jewish life had been lived became a thing of the past. This was not because the decree came like a bolt of lightning, smashing open a closed community and transforming it instantly into a part of the larger society. The cultural integration of the Jews took generations, and they would remain a distinct outgroup well into the twentieth century. The engagement of the Jews with Danish culture, moreover, had been proceeding for a century before the decree appeared. But once the decree was published, this transformation became irreversible. The decree ended the administrative and ritual autonomy of the Jewish community, which had for a century mediated relationships between Jews and Danish society. From 1814 onward, Jews would encounter that society directly, and in doing so they would become part of it.

Integration and Assimilation, 1814—1901[11]

Between 1814 and the beginning of the Eastern European immigration in 1901, the Jews of Denmark experienced perhaps the least dramatic and most unified period of their history. Their movement into Danish society brought a steadily rising level of wealth, a steady expansion of rights, and a gradual decline of the class and ideological divisions that had long fragmented the community. For the first time, Jews became prominent figures in Denmark's literary and intellectual life, and some achieved high positions in government and industry. Some dramas did occur, of course, and some factional quarrels continued. But on the whole, the development of the Jewish community proceeded much as Nathanson and his followers had envisioned it—slowly and steadily, the Jews changed from an outcast cultural isolate to a prosperous and integrated branch of Danish society.

Things didn't start out that way. The 1814 decree did nothing to stem the anti-Jewish feeling that had expressed itself the previous summer, and as the Danish economy continued its decline, accusations that the Jews had sabotaged the nation surfaced repeatedly. In 1819, as the state teetered on the brink of bankruptcy, an outbreak of anti-Semitic violence began in

Germany. On Friday, September 4, it found its way to Denmark. That evening, a mob in central Copenhagen began smashing the windows of Jewish shops and assaulting whatever Jews they could find on the streets (Blüdnikow and Jørgensen 1984: 86–90; Borchsenius 1968: 58–60).[12] Some Jews had to go into hiding, including Nathanson, whose firm had been accused of causing the collapse of the state bank. Agitators put up posters on the Bourse and other buildings, calling on "all good and Christian citizens" to drive the Jews, "that plague upon society, out of the state" (Blüdnikow and Jørgensen 1984: 90). The police who tried to intervene were quickly driven off by the mob. The next day, the crown called in the military, who soon restored order. Similar riots broke out and were put down in several of the provincial cities in the ensuing days. For all its drama, the "bodily Jewish feud" (korporlige jødefejde) caused relatively little property damage and no serious injuries. Nor did it presage more dangerous clashes further on; the crown punished its leaders severely, and with the exception of some youths throwing stones at a Jewish building in 1830, there would be no more anti-Jewish violence in Denmark until World War II. Even so, the riots shook the sense of security of the Danish Jews and made manifest the distance between their growing legal equality and their continuing cultural isolation.

Tumult also occurred within the community early on, as reformers tried to remake religious observance as well as secular status.[13] The 1814 decree had created a new clerical office: that of catechist, the man responsible for instructing the students in the new catechism and conducting the new confirmation ceremony. Nathanson filled the office with Isak Noa Mannheimer, an energetic twenty-two-year-old with strong reformist views. Born in Denmark, Mannheimer spoke Danish fluently, unlike the aging Rabbi Gedalja. He was also a captivating public speaker, who would later become one of the leading spokesmen for the Reform movement in Europe. He began leading classes in 1816, and in 1817 he held the first ceremony. With major reformers and statesmen in attendance, the ceremony embodied the new vision of Jewish ritual. Led by an organ, the audience sang Danish translations of biblical psalms; after that, Mannheimer delivered a riveting sermon in Danish, then examined the candidates. The crowd responded enthusiastically, prompting Mannheimer to seek permission to hold supplemental Danish-language religious services on Wednesday afternoons. The crown approved his request, and on July 9, 1817, he held the first of three such events. They were an immediate success. At a time when most Copenhagen synagogues drew low and unenthusiastic attendance, Mannheimer's sermons packed in 400 or more rapt listeners—even in cramped auditoriums at the height of summer. For Nathanson and the reformist board, Mannheimer represented a new and exciting future for religious observance in Denmark.

For conservatives in the community, however, Mannheimer aroused not so much excitement as horror. His translation of texts, his shortening of the service, his elimination of prayers, his use of music, his mimicking of Protestant style—all amounted to a full-scale assault on Jewish religious tradition. While liturgical differences existed among the various small synagogues in the city, the contrast with Mannheimer reduced such variations to trivia. As a result, traditionalists who had submitted passively to most of the reform agenda raised a unified opposition to the new services. Over 300 of them—including Nathanson's brother—signed a formal protest to the chancery in August. In response, the crown halted Mannheimer's services and appointed a commission to draft a revised format for Jewish observance in Denmark. The commission, which included both traditionalists and reformers, divided so sharply that little agreement was possible. In 1823 Mannheimer gave up the fight and took a rabbinical position in Vienna. The traditionalists returned to their small synagogues, having effectively beaten back the only serious attempt to institute a reform liturgy in Copenhagen.

This tumultuous period reached an end in 1828, when twenty-seven-year-old Abraham Alexander Wolff took over as the congregation's rabbi. Abraham Gedalja had died in 1827 after thirty-six years' service in the community; he led the opposition both to Mannheimer and the 1796 reform proposal, and he had personified the traditionalist Judaism that these movements had tried to overthrow. Wolff, cosmopolitan and academically educated, represented a very different sort of rabbi, one much more attuned to the integrationist sentiments of Nathanson and his supporters. He immediately began organizing the construction of a magnificent new synagogue in Krystalgade. He also developed a new order for synagogue services, which he instituted when the building opened in 1833. This form adopted some reform ideas, such as shortening the services, but it left the vast majority of the traditional service unaltered. Both reformers and conservatives expressed chagrin at the result; a small group of extreme traditionalists went so far as to establish their own small synagogue, which would remain in operation until 1957.[14] Most Jews went along with Wolff's approach, however, and for the first time in a generation the synagogue became a unifying force in the community. Wolff steered a moderate course in other areas as well throughout his long tenure. When he died in 1891, the bitter ideological divisions that had riven the community when he arrived had largely disappeared.

Many other things would also change over the remainder of the nineteenth century. Most of this change occurred incrementally, rather than through the dramatic group confrontations that characterized the reform period. With a few exceptions—the passage of the Danish Constitution in 1849, for example—the community's history during this period contains no momentous dates or major turning points. The change came about, rather, from the common problems of identity and integration that individual Jews faced as they encountered Danish culture firsthand. The history of the nineteenth century therefore shows through more clearly as a set of trends than as a chain of events.

One trend involved the steady expansion of legal rights. As noted above, the Decree of 1814 produced little concrete change in Jewish civil status; while it guaranteed the right of Jews to work, it left in place legal restrictions on their ability to vote, hold office, take oaths, and marry non-Jews. Popular prejudices, moreover, meant that some positions to which Jews had legal access—professorships in the university, for example, or supervisory appointments at hospitals—remained effectively closed off. Over the course of the nineteenth century, virtually all of these legal barriers fell, along with most of the unspoken ones. Jews received the right to vote in the national assembly (*Stænderforsamlingen*) in 1834, and in 1842 a Jew was elected to the Copenhagen city council. In 1849, Denmark's new constitution established religious freedom nationally. The constitution not only assured the Jews of religious legitimacy—along with previously banned groups like the Baptists—but it also lifted the requirements for confirmation services and the royal appointment of rabbis imposed by the 1814 decree. Subsequent years saw the easing of restrictions on the travel of foreign Jews in Denmark (1850), and the moderation of the special oath for Jews (1864). Bars to office fell gradually as well. At Copenhagen University, for example, which had previously barred Jews from teaching positions, Jews gained the right to compete for medical professorships in 1843; by 1872 all formal restrictions on Jewish faculty had been repealed. These successes did not end discrimination against Jews in Denmark, of course, nor did they signal the disappearance of anti-Jewish prejudice in the culture. Even so, as the twentieth century opened, the formal barriers that had separated Jews from full participation in Danish society had largely fallen.

Along with this legal absorption came an increasing cultural one. In 1800, a large proportion of Denmark's Jews had immigrated from other countries, and they lived their lives largely in the context of Jewish institutions. By 1900, the vast majority had been born in Denmark, received their education in Danish schools, and spent much of their lives in non-Jewish social contexts. As Jews thus met the Danish world face to face, they adopted many of its cultural forms. Danish gradually replaced Yiddish as a first language for most Jewish families. Danish clothing styles and hairstyles became increasingly common among Jews, while traditional Jewish practices like kosher housekeeping and Sabbath observance declined. Religious observance appears to have declined for many Jews as well, as they moved from the poor neighborhoods near the synagogue to more affluent ones farther away. Many Jews went so far as to convert to the national church. M. L. Nathanson had all eight of his children baptized in 1828; he changed their last names to "Nansen," and two of them went on to become prominent Lutheran priests. Neither the state nor the Jewish Community kept statistics on the number of conversions that occurred, but anecdotal evidence suggests that substantial numbers did so, especially among the community's wealthier families. Even for those who remained Jewish, the growing engagement with Danish culture strongly influenced perceptions of the self and Jewish identity. At the opening of the century, members of the community clearly saw themselves as Jews; by its close, most described themselves as "Danes of Jewish faith."

This engagement with Danish culture also meant an engagement with Danish society—in some cases, quite literally. Marriages between Jews and Christians increased steadily over the course of the nineteenth century, especially after the introduction of religious freedom in 1849. This trend accelerated sharply as the century neared its end. Between 1880 and 1889, interfaith couples accounted for 35.9 percent of the Jewish marriages in Copenhagen; between 1890 and 1899, they accounted for 40.7 percent, and between 1900 and 1905 they made up 48.2 percent (Blum 1972: 34). Mixed marriages were far more common for Jewish men than women: of the 218 mixed marriages in Copenhagen in 1906, only 71 involved a Jewish woman (Balslev 1932: 92). For both men and women, however, the union usually meant a loss of children to the Jewish community. Of the 370 children born to mixed marriages contracted in 1906, for example, only 61, or about 16 percent, received a Jewish upbringing (Balslev 1932: 92). The trend toward mixed marriage thus had two major social consequences. First, combined with conversions to Christianity, outmarriage kept the Jewish population of Denmark relatively stable in the nineteenth century. Between 1834 and 1901, a period when the Danish population doubled, the number of Jews declined from 4,064 to 3,476 (Balslev 1932: 93). Second, the high rate of intermarriage meant that increasing numbers of Danish Jews had Christian in-laws, grandchildren, nieces, and nephews. Danish society had become intimately linked to Jewish social networks.

Another index of Jewish establishment was Jews' increasing visibility in public culture. Jews were hardly unknown earlier on, of course, as the literary Jewish feud had made clear. They figured occasionally in eighteenth-century art and drama, usually as peripheral characters, and stereotypes of unsavory Jewish peddlers or influential Jewish financiers cropped up regularly in public discourse. In the nineteenth century, however, Jews began to enter the public arena in a different way, as prominent figures in literature, the arts, and public affairs. Writers like Meir Goldschmidt, Georg Brandes, and Paul Levin achieved widespread acclaim in the literary world; painters like Ernst Meyer and David Monies became prominent Danish artists; physicians like Adolph Hannover and Harald Hirschsprung rose to the heights of Danish medicine. Jews helped establish enterprises that became central institutions in Danish culture, including the Bing and Grøndahl porcelain factory, the newspapers *Politiken* and *Berlingske Tidende,* and the banks Privatbanken and Handelsbanken. A few Jews even became prominent politicians, such as Herman Trier, chairman of the Danish parliament (*folketing*) from 1901 to 1905. For many such people, being Jewish posed significant problems. Literary rivals of Georg Brandes disparaged his Jewish origins, while Privatbanken founder David Baruch Adler was forced out by an anti-Semitic director. Some Jews, like the poet Henrik Hertz, therefore converted to Lutheranism and avoided any public mention of their background (Borchsenius 1968: 71–76). Others, however, engaged the question directly—most notably Meir Goldschmidt, whose novels *A Jew* (1968) and *The Raven* (1867) explored the dilemmas of Jewish identity in Danish society. His ability to do so reflected the changing place of Jews in Danish society. The nineteenth century did not merely bring Danish culture into the Jewish community, it also brought Jews—as influential, three-dimensional figures—into the heart of Danish culture.

Jews did diverge from the larger Danish population in one area: their economic standing, which increased over the nineteenth century at a pace far beyond that of Denmark as a whole. At the opening of the century, Denmark's Jewry were largely poor merchants and tradesmen, with a small number of wealthy merchants and a large impoverished underclass. The opening of the trades to Jews in 1788 had begun to change that; the 1814 decree officially opened all other occupations, and over the course of the century Jews found their way into almost all of the nation's most prestigious positions. This process drastically transformed the economic position of the community (Balslev 1932: 88–91). The number of professionals, for example, increased substantially. In 1834 the community included 9 doctors, 3 lawyers, and 5 engaged in writing, journalism, or fine arts; by 1906, while growing only slightly in numbers, it counted 25 doctors, 34 lawyers, and 38 writers or creative artists. The place of commerce remained largely constant (37.4 percent of workers in 1834 and 44.3 percent in 1906), but the scale of operations changed sharply. The number of shopkeepers and peddlers dropped from 204 to 79, while wholesalers rose from 45 to 235; another 55 served as bank directors, stockbrokers, or other high-income merchants. Factory owners rose from 6 in 1834 to 49 in 1906. The result was that by the opening of the twentieth century, most Jews belonged to the upper end of the middle class, with incomes and living standards far above the Danish average.

Taken all together, the nineteenth century embodied both the promise and the danger of the Enlightenment vision of Judaism. Released from legal disenfranchisement and cultural isolation, the Copenhagen Jews experienced a dizzying rise in freedom, welfare, and prestige. They changed from an impoverished and despised outgroup to an integral part of the nation's upper middle class. The bitter ideological divisions that had consumed the community largely fell away, and a large synagogue with an eminent rabbi replaced the scattered dingy apartments in which worship had taken place. At the same time, however, the cohesiveness and tradition of the Jewish community sharply eroded, as its members drifted into the fashions and the families of the surrounding world. In 1914, an article by Rabbi Max Schornstein[15] expressed the dilemma this way: "The danger that a Danish Jew should forget that he is Danish no longer exists, but many are well on the way to forget that they are Jews. At its hundredth anniversary, the Danish congregation, just like the Danish prince, faces the great question: to be, or not to be?" (quoted in Balslev 1932: 92; my translation).

IMMIGRATION, OPPOSITION, AND SCHISM, 1900—1940[16]

This long interlude of tranquility and unity came to an abrupt end in the opening years of the twentieth century. Almost as soon as it began, the new century brought a renewed battle between conservative and liberal factions within the community, one that produced a schism larger than any the congregation had seen before. At the same time, a flood of refugees from Eastern Europe both enlarged and divided the Copenhagen Jews in unprecedented ways. While the forces of assimilation and integration that had characterized the 1800s continued to operate, they were joined by international and congregational dynamics that directed their effects in very different ways.

Some of these dynamics came from within the community itself. After the death of Rabbi Wolff in 1892, his assistant David Simonsen had taken over the office of chief rabbi. Simonsen retired in 1902, and many in the congregation expected his unassuming assistant Hirsch Goiten to replace him. The board, however, wanted a more charismatic figure, and they offered the position instead to a well-known Hamburg rabbi named Tobias Lewenstein. To attract Lewenstein, the board offered him both a lifetime contract and final authority over all religious matters in the congregation. They soon regretted doing so—while Lewenstein was indeed charismatic, he was also rigid and dictatorial, and far more conservative than his predecessors. When Goiten sickened and died in late 1893, rumors attributed his decline to emotional abuse by Lewenstein. The new rabbi had little patience for the compromises with Danish culture that Wolff and Simonsen had worked out over the decades. He objected most fiercely to the congregation's policy on mixed marriages, whereby children of non-Jewish mothers who were raised as Jews were automatically accepted into the community. Lewenstein put an end to this practice. He insisted that such children must formally convert if they were to become Jews, and he refused to promise an easy or automatic conversion process.

Lewenstein's position, while theologically defensible, raised a storm of protest. The congregational board pressed him to relent; when he would not, they took the extraordinary step of hiring a second, independent rabbi for the congregation—Dr. Max Schornstein, a young rabbi from Bohemia with less conservative views. Lewenstein regarded this action as a breach of his contract, and he took equally extraordinary steps to counteract it. He took out a newspaper advertisement proclaiming that no admittance to the community was valid without his approval; he also took the Jewish Community to court, seeking damages for breach of contract. A period of virtual open warfare followed, eagerly covered by the national press. The two rabbis alternated sermons in the synagogue, and each used his time to argue obliquely for his position. At times the bitterness of the feud approached the comic. Whenever Lewenstein told a fable involving a lion and a donkey, for example, he would cast himself as the lion and point directly to Schornstein when the donkey appeared (Melchior 1965: 42). The board eventually fired Lewenstein, an action he refused to accept; they removed his chair from the synagogue, but he came defiantly into services anyway and stood at the front in his usual place.

In the end, Lewenstein won his lawsuit. The Jewish Community paid him damages of 85,000 kroner, and Schornstein took over as chief rabbi. Lewenstein did not leave the country; a circle of supporters, many of them youths or Eastern European immigrants, set up an alternative synagogue under his leadership. Named Machsike Hadas, or "Bearers of the Myrtle,"[17] the group was initially intended merely as a moral support for Lewenstein. When he left Denmark in 1912, however, its members' hardening antipathy for the larger congregation made a return there impossible. With funding from the international Orthodox Jewish association, they managed to hire a successor and establish a permanent synagogue. Most of its members remained formal affiliates of the Jewish Community, which still administered the nation's only functioning Jewish cemetery. They set up a number of alternative institutions, though, including a religious school, a ritual bath, and a kosher delicatessen. For the remainder of the twentieth century, Machsike Hadas would remain a significant presence among the Copenhagen Jews, a vocal conservative force in an increasingly liberal congregation.

More profound changes in the Jewish community came from a very different source: the wave of pogroms and other anti-Semitic violence in Eastern Europe. Beginning with the 1903 pogrom in Kishinev, these events drove hundreds of thousands of Jews westward from the Russian Empire. Many passed through Denmark on their way, and some of them decided to stay. In 1903, 264 refugees arrived, prompted by the atrocities in Kishinev. By 1911, there were 1,600, and by 1916, 2,443 (Blum 1972: 37). In 1917, a royal decree officially closed off further immigration, but the outbreak of World War I inevitably increased the number of refugees. These events transformed the demographics of Copenhagen Jewry. The population, which had declined from 3,264 in 1890 to 2,826 in 1900, shot up to 5,875 by 1921; of that number, 3,146 were either immigrants or their children (Blum 1972: 37). In the course of two decades, the Copenhagen Jews changed from a declining and largely homogeneous middle-class community to a large, diverse, and sharply polarized one.

These new arrivals represented something profoundly new for the Copenhagen Jews, a sort of cultural contrast they had not encountered before. The Jews had, of course, experienced divisions in the eighteenth century, both among classes and among ideologies, but these differences had presupposed a broad cultural commonality. For the most part, despite their oppositions, the Jews had shared a language, a religious tradition, and a culture drawn from a limited area of northern Germany.[18] The new immigrants—known colloquially as "the Russians" (*russerne*), in contrast to the native "Viking Jews" (*vikingjøder*)—departed sharply from each of these (Blüdnikow 1986). Where Danish had become the first language for the Vikings, most of the Russians spoke Yiddish, using Eastern European dialects and mannerisms. Where the Vikings had come to look increasingly like the Danes, the Russians looked like Eastern Jews, with dark clothing, fur hats, and large beards. Where the Vikings' religiosity had veered toward liturgical reform and lax observance, many of the Russians practiced strict religious observance drawing on the Hasidic movement. Where the Vikings had become increasingly associated with political conservatism, moreover, the Russians included many Bundists and socialists who had fled Eastern Europe for political reasons. The huge influx of these Jews, who quickly became a majority in the congregation, demanded the rethinking of the meaning of Jewish identity in Denmark.

The physical visibility of the Russians had implications in itself. The refugees who settled in Denmark did so largely out of poverty—many would have preferred to emigrate to America, but they lacked the money for a transatlantic passage. The vast majority worked in the garment trades as tailors or seamstresses. They had little formal schooling and no knowledge of Danish; as a result, the only work available to them was grueling sweatshop labor, often for Jewish garment manufacturers in Copenhagen.[19] While such labor could provide a bare subsistence for a small family, it offered little security, and an illness or sudden expense could easily push a family into ruin. Families were not, moreover, small; the Russians had many more children than the Viking Jews, stretching their meager resources even further. The Russians therefore lived in the poorest slums of the central city, forming a teeming Jewish "ghetto" of a sort never seen in Copenhagen before. Dark-skinned and dark-haired by Danish standards, their adults in foreign garb and their flocks of children in tatters, talking loudly in a foreign language, the new Jews stood out dramatically on the streets of the city. They made a particularly strong impression on Saturdays, when large numbers would walk in groups to Sabbath services at the synagogue. After a century of steady acculturation, during which Jews had become virtually invisible in Danish culture, the Russians presented a radically different picture of Judaism to the Danish public.

Opinions differed among the Viking Jews in how to respond to this change. As early as 1904, a group of Jews formed the Russian Jewry Assistance Committee, and in 1912 the newly established B'nai B'rith became a major source of support for the immigrants. These organizations raised money both within the community and abroad to help the Russians. In 1908 alone, they distributed over 75,000 kroner to refugees. Their members disagreed, however, about what form this help should take. Many of the Viking Jews wanted to send the Russians elsewhere as soon as possible, and the Jewish Community's policy for the first few years stressed this objective. Until 1910, the largest expenditures went to pay for tickets to the United States, England, or other parts of Scandinavia. The administrators abandoned that objective after 1910, as it appeared hopeless, and they focused their efforts on assistance with housing, medicine, and food. The Jewish Community had long maintained independent poor relief programs, which were extended to the Russians, and the Jewish schools were opened immediately to the immigrants' children. Special programs helped to buy ritual foods at Passover and wedding dresses for brides. Some leading Jews, notably Rabbi Schornstein and the former rabbi David Simonsen, acted as advocates for the Russians both to the government and to the national press. At the same time, though, the Jewish Community presented a dictatorial and often scornful face to the beneficiaries of its support. Russians who applied for assistance lost their rights to vote in the Community; they were put on a "beggar list," which also counted against them if they applied for Danish citizenship. Accordingly, even after the Russians had become a numerical majority among the Danish Jews, the Vikings continued to hold almost all Community offices. The head of the relief program had absolute authority over who would and would not receive help, decisions that many Russians saw as arbitrary and capricious. Some "problem cases" were reported to the government for deportation. The Community's efforts did provide substantial help to the poorest immigrants. By humiliating and disenfranchising their subjects, though, the programs also planted lasting seeds of resentment (Blüdnikow 1986).

The result was a deep and bitter polarization in the Jewish community, one that lasted for decades. Russians seethed at the Vikings' superiority; according to a common saying, "They don't even deign to spit on us" (Melchior 1965: 31). The Vikings, meanwhile, chafed at the Russians' public image, which seemed to undo a century of increasing Danishness and social integration. "People used to see us as Danes of the Mosaic faith," complained one speaker at a B'nai B'rith forum on the issue. "Now they see us as Jews!" (Blum 1972: 38).

Over time, some of these oppositions began to erode. The Russians established themselves very quickly, in large part because of their children's early entry into the Jewish schools. By 1930, they had begun to move out of the "ghettos," and most of them had achieved some financial stability. As their dependence on Jewish Community support declined, they developed a number of independent organizations, including Yiddish newspapers, drama societies, political clubs, and choirs. Over time, however, Yiddish organizations gave way to an increasing immersion in Danish culture. Most of the Russians had come from fiercely anti-Semitic societies, and their relatively easy acceptance by Christian Danes came as a shock. Memoirs from the period include repeated references to their disbelief at seeing their children playing peacefully with blond Christian neighbors.[20] Indeed, many Russians found the Danish Christians considerably more appealing than the Danish Jews. By the 1930s, intermarriage rates among the Russians had begun to approach those among the Vikings, and Danish had largely displaced Yiddish as a first language among the young. These trends tended to diminish the antagonism between the Vikings and the Russians, as did the increasing number of marriages and business relationships between them. As the decade came to an end, the Vikings were still wealthier and more Danicized than the Russians, but—especially in the face of events to the south—they saw themselves increasingly as a single community.

Occupation, Exile, and Return, 1940—1945 [21]

The period of World War II constitutes a critical chapter in the history of the Danish Jews, one that we will discuss at length in chapter 7. Here, let us merely sketch the basic events of the time, to place the later history of the community in context. For the fate of the Jews under Nazi occupation—their protection, rescue, and later return—dramatically transformed both their community and their place in Danish culture after 1945.

The Jews' fortunate position during the war derived in large part from the political circumstances under which Denmark came into the conflict. On April 9, 1940, German forces occupied the country, supposedly to guard against a British invasion through Jutland. Germany did not, however, treat Denmark formally as a conquered country. Hitler wanted Denmark to serve as a "model protectorate," an example of the benevolent world order that Germany planned to establish. The Danes were therefore to maintain their own civil government, their own constitution, and their own administration, with German control limited to foreign policy. This plan aroused considerable resentment among the Danish public but very little active resistance. Attempts to build Nazi organizations in the country met with very little success, and such symbolic resistance as King Christian's horseback rides through Copenhagen found strong public support. At the same time, though, the government acquiesced to German orders, the police helped root out resistance cells, and Danish farmers provided abundant crops to the German war machine. Denmark's overall policy aimed to minimize the impact of an occupation it could not prevent; Winston Churchill derided the nation as "Hitler's canary."

The Danes did resist German demands on one notable point: the government refused, despite repeated German pressure, to take any measures against the Jews. Leaders insisted that the Danish Constitution protected religious minorities, and that any anti-Jewish actions would violate German promises to respect that document.[22] They extended this argument to prohibit even a census of the Jewish population, which could have been used to target Jews for mistreatment. As a result, after a few weeks of anxiety, most of the Jews found their status virtually unchanged under the occupation. They kept their jobs, the synagogue remained active, and few found cause to leave the country. Indeed, as the Jews became a sort of proxy for national independence, their image in public culture improved. Anti-Semitism had never been strong in Denmark, but during the occupation it disappeared almost entirely. Leaders like Hal Koch called loudly for the Jews' protection, and King Christian X sent public messages of support.[23] As the Holocaust raged through Europe in 1942 and 1943, Denmark's Jews lived on largely undisturbed.

Things changed in 1943, a year when Danish resistance to the occupation turned increasingly violent. Relations between Danish and German authorities deteriorated during the summer of 1943, and on August 29 the protectorate arrangement collapsed. The government resigned, the country was placed under martial law, and German plenipotentiary Werner Best assumed dictatorial control of Denmark. Free at last to act, Best quickly began laying plans for the roundup and deportation of the Jews.[24] Using a list of addresses stolen from the Jewish Community offices, Gestapo officers made plans to surround Jewish homes on the evening of October 1. On this, the evening of the Jewish New Year, Jewish families would be gathered in their homes for holiday dinners; they would be seized, whisked onto boats waiting in the harbor, and out of the country before daybreak.

Best anticipated little resistance and at most a brief wave of unrest when the news of the roundups broke.

He could hardly have been more wrong. Word of the operation leaked out a day early, and a massive effort to warn and hide the city's Jews began. Jews found refuge with Christian friends, neighbors, and employers; many were hidden in the city hospital, checked in under false Christian names. Some of this effort was organized by the Danish resistance, but much of it was spontaneous, as Jews found themselves approached in streets and workplaces by strangers offering shelter. When the Gestapo squads arrived at Jewish homes, therefore, they found most of them empty. Over the next two weeks, the vast majority of the country's Jews were ferried in fishing boats to neutral Sweden, using an elaborate system of way stations and transport links set up by the resistance. Of about seven thousand Jews living in Denmark in 1943, the Germans succeeded in capturing only 472; these they sent to Theresienstadt, where 51 of them died of disease and malnutrition. Because of persistent agitation by the Danish Red Cross, however, none of the Danish Jews were sent further to the death camps, and the survivors were allowed to leave the camps for Sweden before the end of the war.

The Jewish experience in Denmark during the war certainly had its traumas—even for the majority who escaped the roundups, the two years of exile in Sweden involved real hardships, both emotional and financial. By comparison with the fate of most European Jews, however, the Danish situation seemed impossibly mild. Over 95 percent survived the Nazis, compared to only a third of the Jews in occupied Europe as a whole. Moreover, unlike other European Jewish survivors, the Danish Jews emerged from the ordeal with a profoundly enhanced sense of connection to the larger society. For most Jews, the events of 1940–1945 demonstrated with appalling clarity their exclusion from the cultures within which they lived; in Denmark, it provided a dramatic demonstration of their inclusion. The symbolic expressions of this inclusion were often extraordinarily moving. To cite one example among many: in their haste to escape, the Jews had left the Great Synagogue unprotected, and many expected to find it in ruins on their return. Particularly mourned was the loss of the Torah scrolls, the sacred parchment texts that would surely have been burned or defaced by the Nazis. On their return, however, to their considerable shock, the rabbis found the scrolls intact. Priests from a nearby church had broken into the synagogue on the night of the roundups, removed the scrolls, and hidden them in a church basement until the end of the occupation. Such actions expressed a concern not merely to save the lives of the Jews but to safeguard their culture as well. As the horrors of the Holocaust came to light in the months and years after the liberation, this concern seemed little short of miraculous.

AFTER THE RESCUE, 1945—2000

One way to measure the magnitude of the 1943 rescue in Danish Jewish consciousness is to read the historical work written on the community. For the most part, it simply stops there; while Jews have written abundantly on the community's earlier history, they have published remarkably little on the events that followed. Bamberger's 1983 history, for example, ends with the joyous return of the refugees from Sweden in 1945. Likewise, a prominent museum exhibition on "The History of the Jews in Denmark 1622–1983," mounted at the Jewish Museum in New York in 1983, contained neither a word nor an exhibit from the period after the war (Barfod et al. 1983).[25] Part of this neglect reflects interest in the rescue, which has tended to push other aspects of the community's history into the background. It also reflects, however, a sense that the rescue marked a new era for Danish Jewry, one to which history, as the study of the old ways, doesn't seem to apply. It does, of course; the postwar period has manifested many of the same trends and tensions that marked the community in earlier years. But in discussing the community's past with its members, one gets an unmistakable sense that 1943 was a dividing line, an unusually definitive point at which the past ends and the contemporary world begins.

We will explore that world in more depth in the chapters to come. Here, we will note only a few events and trends that have characterized the community in the years since the war.

The first years after the war were tumultuous ones. The return from exile in 1945 created a cascade of intense and conflicting experiences. Some Jews returned to find their apartments cleaned and maintained, with meals and flowers set out by attentive Christian neighbors to welcome them; others found their homes destroyed or occupied, their property stolen, their jobs gone. In the frenzy of recrimination and self-examination that followed the war, many Danes celebrated the Jews as evidence of the nation's anti-fascism and humanitarianism; at the same time, the confusion and disorientation of the period produced a brief but palpable wave of anti-Semitism. A substantial number of Jews, worn out by their persecution and exile, said that they were "tired of being Jewish" and converted to the state church; others experienced a renewal of Jewish consciousness in the Zionist movement or in Jewish youth organizations. For the Danish Jews, as for Jews around the world, the immediate postwar period involved confusion and conflict, as Jews tried in different ways to assimilate the momentous implications of the war, the Holocaust, and the creation of Israel.

By the mid-1950s, matters had settled out considerably, and a sense of a new shape for Jewish life in Denmark had begun to emerge. Some aspects of this shape were distinctly new. The state of Israel, for example, assumed a central place in both Jewish self-perception and the organization of the Jewish Community. By and large, the community had distanced itself from Zionism before the war; afterward, however, Danish Jews embraced the cause with abandon. A number of young men, including the future chief rabbi Bent Melchior, traveled to Palestine in 1948 to fight for Israeli independence. Others spent time living on kibbutzim or even emigrated to the new country altogether. Back in Denmark, Zionist fundraising societies became some of the most important voluntary organizations in the Jewish Community. This trend constituted a real change for Danish Jewry, which had become steadily more Danicized and provincial over the preceding century. As the Jewish Community committed itself to Zionism, it encouraged its members to link their Jewish identity with something external to Denmark, something that distinguished them as a group from their non-Jewish countrymen—a reversal of the Community's position since the days of M. L. Nathanson.

Indeed, a general movement toward fostering a distinctive Jewish identity developed in the postwar period. Jewish organizations had existed before the war, of course, and many of them had promoted Jewish culture, but that had rarely been their express purpose; the Yiddish theater groups, for example, reflected interest in seeing Yiddish plays, not a concern to maintain Yiddish culture. In the postwar era, maintenance of Jewish identity became a critical concern. The Jewish schools, which had been established to ease Jews into mainstream Danish culture, reversed their role completely. By the 1970s, parents sent their children to the Carolineskole not to learn to be Danes, but to learn to be Jews. Youth clubs, cultural groups, sports associations, and other institutions increasingly justified their activities as mechanisms for the promotion of Jewish consciousness. In the decades after the war, the project of perpetuating Jewish identity—and the gnawing fear that it could not be done—moved to the center of organizational activity in the congregation.

This concern with the precariousness of Jewish identity indicates one of the continuities between the prewar and postwar periods: the rapid pace of acculturation and assimilation into Danish culture. The rescue produced several decades of pronounced philo-Semitism in Denmark, as the Danes' heroic role became a key ingredient in the reconceptualization of national identity after the occupation. Anti-Semitism virtually disappeared from everyday life, and the social divisions and prejudices that had separated Christians and Jews largely evaporated. As a result, the intermarriage rate continued to climb dramatically. By the 1990s, only a small minority of the country's Jews married Jewish spouses. Jews moved more and more into positions of national prominence, including leading political positions; some, like Victor Borge, became international representatives of Danish culture.

Another continuity was the tension between immigrant and inborn Jews in Denmark. The war effectively ended the split between the Vikings and Russians; having found themselves quite literally in the same boat during the rescue, the two groups mended fences upon their return. Between 1969 and 1971, however, a new group of refugees arrived, about 1,500 Jews fleeing persecution in communist Poland. These new arrivals differed sharply from their predecessors. Most of the Poles were urban, educated professionals, often with little religious involvement. Many were committed communists, and some had held important positions in the party bureaucracy. For all their differences, though, their relationship with the Jewish Community echoed many themes from earlier in the century. While they received a warm welcome from Danes in general, many Poles found their reception by the Jews cold and scornful. Danish Jews referred to them as "Polacks," snickered at their poor Danish and their strange manners, and treated them as second-class citizens in Jewish Community affairs. Like the Russians, the Poles formed separate organizations and social circles, some of which are active today. Like the Russians' children, theirs have acculturated rapidly, and their integration has softened some of the antagonisms between the established and immigrant groups. Despite this rapprochement, immigration still provides a point of tension in the community, and more recent immigrants—including a substantial number from the Middle East in the 1990s—continue to report exclusion and prejudice.

The role of the chief rabbi has also remained a point of tension in the Jewish Community, sparking some of the most energetic political disputes in the postwar period. Marcus Melchior, a member of one of the old Viking Jewish families, set the tone for the position when he succeeded Max Friediger in 1947. A popular and highly visible public speaker before his appointment, Melchior dominated the community through force of personality until his retirement in 1969; while his formal authority extended only to religious issues, his popularity gave him considerable influence over the Jewish Community's political life as well. This influence brought the rabbi's role into conflict with that of the Community Board, which held formal authority over congregational affairs. When Melchoir's son, Bent, succeeded him in 1970, this tension broke into the open. Like his father, Bent Melchior used the rabbinate as a political platform, becoming a prominent spokesman for liberal causes in Denmark and Europe at large. Chafing at his influence, the board attempted to remove him twice, in the early 1970s and in 1981–1982. Neither effort succeeded, as Melchior managed to muster enough support in the congregation to head off his dismissal. Indeed, in 1982, he managed to unseat most of the board in the process (see chapter 4). The result has been a weakening of the congregational board, which has seen its influence and its prestige decline steadily during the postwar period. More and more, the chief rabbi's personal and religious authority have eclipsed the formal structures of Jewish Community governance.

A final familiar theme in postwar Jewish life has been an enduring polarization between liberal and conservative wings of the congregation. As Jews have integrated more and more into mainstream Danish society, they have increasingly taken on mainstream Danish attitudes toward religious observance. Like Christian Danes, most have tended to associate religiosity with the old, the irrational, and the anti-modern. While they admire the perceived authenticity of the deeply religious, they have generally moved away from traditional observance. The religious, meanwhile, have become increasingly isolated during the postwar period. Oppositions between the large congregation and Machsike Hadas have deepened; for many in the orthodox wing, connections to orthodox groups in Israel and England have become more important than connections to the Danish Jews. This opposition has old roots in Copenhagen, of course, but it has taken several new turns in the years since 1945. One is a shrinking of the conservative wing, as many of its members have emigrated to Israel or England. Machsike Hadas has dwindled to a small circle of related families, its activities supported largely by the patronage of a single wealthy member. Another is a growing political boldness among members of mixed marriages, who have pushed increasingly for the formal inclusion of their spouses into the Jewish Community. In recent decades they have achieved a new visibility in the community, with one of their number, Jacques Blum, becoming Jewish Community president in 2000. Perhaps the most intriguing development is one of the most recent: the arrival in 1996 of a pair of Chabad missionaries, who have established a small but active Hasidic presence in the congregation. With the simultaneous prominence of both Blum and Chabad, the polarities that have long characterized Danish Jewry—between liberalism and conservatism, between religiosity and cultural Judaism, between an ideal of Danish integration and an ideal of Jewish distinctiveness—stand out more sharply than ever.

Conclusion: History, Identity, and Community Among Danish Jews

The more things change, the more they stay the same. It is tempting, looking at the history of the Danish Jews, to come to such a conclusion. In 1684, the community was divided between Ashkenazim and Sephardim; in 1784, it was divided between reformers and traditionalists; in 1984, it was divided between liberals and conservatives. At the turn of the eighteenth century, established Jews resented the damage to their reputation caused by the immigration of poor Jewish peddlers; at the turn of the twentieth, Viking Jews resented the new image of Judaism brought in by the Russians; near the turn of the twenty-first, Danish Jews looked askance at the otherness of Poles and Middle Eastern immigrants. Factionalism, religious disagreement, and conflicts over immigration and occupation have all maintained a steady presence throughout the Jewish experience in Denmark. Above all, at least since the 1700s, the Danish Jews have struggled with the central question that pursues any diaspora community: how and how much should one seek to engage with the larger society, and how and where should one buttress one's own distinctive group identity? Not only have such questions remained constant, but in many cases their answers have kept much the same form. M. L. Nathanson and Rabbi Gedalja, for example, faced vastly different times than do Jacques Blum and the current chief rabbi, Bent Lexner; two hundred years of history, including an industrial revolution and two world wars, stand between them. Yet each Jewish Community president could argue for intermarriage in almost the same words, and each rabbi would phrase his rebuttal almost identically. For all the transformations that the world has undergone since the seventeenth century, much about the Jewish condition in Denmark has stayed the same.

We should be careful, however, not to let this observation blind us to the real changes that have occurred. However many parallels we can draw between Jewish worlds in 1684 and 1984, they were not the same worlds; their members spoke different languages, held different occupations, practiced different styles of religious observance, and had radically different relationships with their neighbors. While they did face many of the same questions, they answered them against very different social and cultural backgrounds. Because of the patterns of immigration and assimilation, even genealogical relationships between the present and the past are limited. Indeed, one might plausibly ask whether the Jewish community today has any significant connection with the one of three hundred years earlier, beyond a common religious tradition and a few attenuated kinship links. To speak of Danish Jewry as a continuous historical entity is at least theoretically questionable; to speak of it as fundamentally unaltered would be absurd.

Among the many things that have changed since the Jews' arrival in Denmark, two turning points have particular significance. The Decree of 1814 fundamentally transformed the nature of the Jewish Community; it turned the Jews from an autonomous and alien entity to an integrated element of the Danish state. The Jews had been in contact with Danish culture before that, but Jews as a group had always stood outside of mainstream Danish culture. The dissolution of their institutional autonomy in 1814 changed the basic nature of Jewish identity in Denmark, making what had been a question of group membership into a matter of individual personal affiliation. In doing so, it removed much of the group's authority to establish who was and was not Jewish, leaving it possible for individuals to come to widely different conclusions about their own status. In dissolving the categorical definition of Jewishness implicit in its institutional structures, the decree made Jews into something fundamentally different.

The other key turning point was the wave of Eastern European immigration that began in 1901. Before this influx, the world of Danish Jewry had been relatively isolated; outside of contacts with northern Germany, from which most Jewish Community families had emigrated, Copenhagen's Jews had relatively limited interaction with the world beyond Denmark. This parochialism deepened during the nineteenth century, as Jews increasingly identified with Danish culture and society. The flood of immigrants at the opening of the twentieth century ended that isolation, intruding the culture and politics of Eastern Europe irrevocably into the community. Subsequent waves of immigration—after World War II, during the Polish persecutions, during the 1990s—continued the process, so that at every point during the twentieth century, recent immigrants from very different cultures made up a significant portion of Danish Jewry. Along with these immigrants came international ties, relationships to social and kinship groups far beyond Danish borders. The establishment of the state of Israel, moreover, gave Danish Jews a new international reference point for Jewish identity. As Danish Jews work out notions of identity and religiosity today, they look at least as much to Israeli ideas and fashions as to those of the Mosaiske Troessamfund; many also have close family members living in Israel (see chapter 5). While being Jewish has always involved a connection with international Jewry, the twentieth century has seen a deepening and strengthening of that connection.

As the Jews of Denmark enter their fourth century, therefore, they face questions about the meaning of the Jewish community that would have been unthinkable to the partisans of the Wig Wars. When Wulf Wallich wanted to change the face of Judaism in the 1780s, he knew what to do. He had a visible authority to challenge, a central place to confront it, and a clear symbolic language in which to express his views. Today, he would find the job considerably harder. The rabbis and other Jewish leaders no longer define the nature and obligations of Jewishness in Copenhagen; more and more, that power has devolved onto individual Danish Jews and to Jewish cultural centers overseas. The state, meanwhile, has appropriated the power over daily life once centralized in the Jewish leadership. Not only does this imply a weaker role for Jewish authorities, it makes it hard to say just what the community is and where its boundaries lie. Today's Jews face the problem of how to make a meaningful community without the kinds of authority, definition, and common understandings that traditionally held them together. They have managed to do so, and to do it surprisingly well. In the chapters that follow, we will explore how that has happened.

2

The Religious World

Faith and Ritual Practice
in Jewish Copenhagen

September 29, 1998, the first evening of Yom Kippur. I am sitting in a prime seat in the Great Synagogue of Copenhagen, five rows from the front and close to the center. The seat is not comfortable; like all those in the synagogue, it is part of a narrow wooden pew, with a high straight back that digs into my spine. But the view is magnificent. To my right stands a large raised platform, the *bimah*, where a cantor in black robes sings prayers in Hebrew from a massive *Torah* scroll. Before me, in a large recess in the eastern wall, two massive candelabras flank the twelve-foot doors of the Torah ark. I can see the silk and velvet draperies of other scrolls within it. All around me, the architecture of the enormous room evokes the baroque grandeur of the last century—gilded latticework on the front wall, massive gilded columns holding up the balcony, ornate brass railings and chandeliers, an intricate carved ceiling. And for once, the activities on the floor seem to match that grandeur. Attendance at the synagogue is sometimes sparse, but today the pews teem with men in white skullcaps and prayer shawls. Many of them are bent over battered *siddurs*, mouthing the prayers along with the cantor; some talk to their neighbors, or wave to friends in other pews; some, like me, sit in silence, gazing around them, taking in the rare majesty of the Day of Atonement. To my left, a bearded man of middle age stands as he prays, bowing rhythmically with the Hebrew words.

Just above the ark, carved into a rectangular recess, an inscription in golden letters spans a quarter of the width of the synagogue. I have often wondered what it says; I don't read Hebrew, and I have never had the occasion to ask anyone about it. Today, however, surrounded by four hundred men praying in Hebrew, it occurs to me that I should be able to find out. I turn to the man next to me, a cheerful soul in his thirties who is bent over a prayer book, and ask him what it says. Surprised at the question, he puts on a pair of glasses and peers at the inscription.

"Sh . . . viti . . . adonai . . . la . . . negdi . . . tamid . . ."

He fumbles with the text for a moment, then grins sheepishly at me. "Actually, I don't really speak Hebrew," he says. "I can sound it out and sing along with some of the prayers, but I don't know what it means. But don't worry," he adds quickly, "my dad's here. He's come here for fifty years, he can tell us." He turns to a venerable-looking man next to him, also bent studiously over a *siddur,* and speaks in his ear. The father, like the son, pulls out a pair of glasses and looks intently at the inscription.

"Sh . . . viti . . . adonai . . . la . . . negdi . . ."

After a moment he too gives up, and shrugs his shoulders with a smile. His son is crestfallen, but also amused, and tells me not to despair—he points up at the bearded man next to me, still intoning the text along with the cantor. I try timidly to get his attention, without success; my neighbor takes a more direct approach, tugging at his coat sleeve and then pointing up at the inscription as he asks the question.

"I can't see it from here," he says.

"No, no, I mean the big one, right there," I say, standing up and pointing at it. Our heads are about a foot apart.

"I'm sorry, I just can't see it from where I'm standing. Maybe after the service." He looks down at his book and resumes his prayer, ostentatiously ignoring me. My other neighbor chuckles and tries the man behind him, while his father gestures across the aisle to a friend.

Over the next five minutes, the three of us ask almost everyone within easy hailing distance, about ten people altogether; none can shed any light on the meaning of the inscription, although several make guesses about particular words. We finally give up, as a new phase of the service begins, and I move to a different part of the synagogue. I will not learn the meaning of the inscription until months later, when I am back in the United States, and I take the easy route of asking my uncle.[1]

This episode raises a simple question: what were these men doing in the synagogue? Anthropologists generally link ritual participation either to personal meaning or to social solidarity—people join in rituals either because the liturgies express meaningful things to them, or because the actions of the ritual express their close ties with the assembled community. In this case, however, much of the liturgy clearly had little direct semantic meaning for most of the people in attendance. Unable to read the verse on the wall, they could surely not understand the elaborate Hebrew of the Kol Nidre service. To the extent that they did understand it, moreover, very few of them believed in it. Most of the men around me that day were middle-class Danes, raised in the nation's secular culture and educated in its secular schools; they no more believed in the premises of the service than they believed that elves and trolls inhabited the Danish countryside. Nor was this a particularly close or solidary community. None of the men I asked about the verse knew their neighbors well enough to realize that they did not understand Hebrew. And indeed, the Danish Jews constitute a notoriously fractious group, riven with bitter social and doctrinal antagonisms. In many ways they embody precisely the kind of fragmented, cosmopolitan, postmodern community in which religion is supposed to die.

Yet the Jews of Copenhagen have a surprisingly active religious life. The synagogue almost always draws at least a hundred or so worshippers on Saturday mornings—a small percentage of the Jewish population, but a number that dwarfs attendance at most of the city's Lutheran churches. Attendance runs much higher at holiday rituals like Yom Kippur services, when Jews pack the synagogue to standing room for services that can last a full day. Rituals like funerals and circumcisions hold tremendous significance for most Danish Jews, while customs like dietary law and Sabbath observance set even private home life in a religious context. More than that, religion stands at the heart of how Copenhagen Jews imagine themselves as a community. To be Jewish means different things to different people, but for almost everyone, religion is one of its basic ingredients. In my interviews, even Jews who never attended the synagogue spoke of it as central to Jewish life, and they measured their own closeness to the community at least partly by their participation in Jewish ritual. Whatever it is to be Jewish in Denmark, the events of the synagogue and other ritual activities are a central part of it.

In this chapter, we will examine why that is. We will discuss the Jews of Copenhagen as a religious community, looking at some of the beliefs that they hold and the ritual activities in which they engage. We will ask what those activities mean to them—why it is that in spite of their lack of belief and their fragmented community, religious ritual continues to have so central a place in the Jewish world. We will not find a single answer. The religious practice of the Jewish community includes a constellation of ritual forms, each of which carries a distinct kind of significance to differing groups within the Jewish community. What they all have in common is their ability to help Jews express and mediate ideas about the self and about the nature of Jewishness. They provide symbolic vehicles through which Jews can make sense of who they are, and settings in which their different understandings of self can come together and interrelate. The religious world of Jewish Copenhagen neither reflects nor creates a uniform or solidary community; it creates, rather, an arena of symbolic action, a set of spaces and symbols with generally understood meanings, which individuals can use in their constructions of themselves and each other. In the confusion of late modern Copenhagen, amid the fragmentation and instability of the Jewish world, such a tool is precious indeed.

BELIEF

Let us begin with the question of faith. What do Danish Jews believe about the nature of the divine, the meaning of human existence, and the moral foundations of the human community? Jews have never had an easy time answering such questions; disputes about religious principles have divided Jews for millennia, finding their way into the religion's sacred texts as well as daily discourse. Most Jewish theologians agree on a number of basic principles, such as the unity of God, the divine origins of Jewish law, and the election of the Jews as God's chosen people. The Jewish scriptures also provide a common reference point, offering a distinctive set of origin myths and historical events commonly understood as the foundations of Jewish religion. No consensus has ever existed, however, on how to interpret these foundations. Should the stories in the Bible be understood as literal history, meaningful imagery, or primitive superstition? Should the personified God of scripture be seen as a literary convention, a tribal belief, or the true image of the Almighty? To what extent is Jewish history rooted in a divine plan? To what extent does Jewish faith incur moral obligations, and what might those obligations be? Such questions have spawned factional and sectarian movements throughout recorded Jewish history, and they continue to do so today. As individual Jews in Copenhagen work out their own understandings of the divine, their religious heritage offers a bewildering variety of options.

The larger Danish culture, meanwhile, offers a powerful suggestion of its own. Most of the Danish Jews live in urban Copenhagen, a setting with a pronounced antipathy to traditional religious belief. While the overwhelming majority of Danes belong to the state Lutheran church, only a tiny minority attend religious services on any regular basis; few avow a firm belief in a well-defined God, and most look to science rather than religion to make sense of their world (Buckser 1996a; Salamonsen 1975).[2] Indeed, over the past century, Denmark has become one of the sociology of religion's type cases in the secularization of modern society (see, for example, Hamberg and Pettersson 1994). Danish popular culture associates religiosity with narrow-minded traditionalism. In literature, film, and journalism, the religiously devout figure primarily as obstacles to progress—Islamic dogmatists frustrating the assimilation of Turkish immigrants, perhaps, or Inner Mission Lutherans protesting the ordination of women. In a culture that prizes modernity and progress, religious belief carries a taint of the past that most urban Danes find uncomfortable. To the extent that they immerse themselves in Copenhagen culture—as most of them do—Danish Jews experience a powerful pressure to distance themselves from religious belief.

Caught between the eclecticism of Jewish theology and the secularism of Danish culture, individual Jews navigate questions of belief largely by themselves. No community consensus exists on matters of belief, nor is any position represented as the official viewpoint. When the Mosaiske Troessamfund published a booklet about itself in 1996, it included ten pages on history and three on ritual practice, but not a sentence on religious beliefs (Dessauer et al. 1996). Similarly, in discussions about community divisions, my informants almost never brought up theology as the source of disagreements. While different ideas about ritual practice may logically presuppose different theological viewpoints, those viewpoints almost never appear in discussions about practice. Indeed, my informants—even very orthodox ones—almost never mentioned their beliefs unless I asked about them directly, and then they talked of them briefly and obliquely. For Copenhagen Jews, religious belief is essentially a private affair, seldom discussed in public and never discussed as a community.

When they did talk about belief, most Jews I spoke with echoed viewpoints common among Copenhageners generally. They gave some credence to the historical dimensions of the Jewish scriptures, and they often posited a somewhat vague belief in God. They regarded the details of the biblical narrative, however, particularly those involving God's interaction with human beings, with a profound skepticism. The notion that God had created the world, that He had personally given the law to Moses, that he had played the roles ascribed to him in the stories of the flood or the sacrifice of Isaac, struck most of my informants as a kind of primitive mythology. In many cases, this rejection of traditional religion seemed to comprise an informant's entire theological universe. While a few had developed sophisticated theological vocabularies, more often my informants found it difficult to elucidate their own beliefs, and they seemed not to have given the subject a great deal of thought. This approach to religion fits well with the secular orientation of Danish society, and it underlines the engagement of most Danish Jews with the larger culture. Jewishness for them does not imply a difference in belief so much as in perspective, a different standpoint from which to embrace a common secular worldview. What makes their faith Jewish is its point of departure—most Jews don't believe in the Old Testament, whereas most Christians don't believe in the new one.

To the extent that individuals do believe in the Bible, the secular orientation of the culture keeps such beliefs out of public discourse. One of my first Sabbath dinners in Denmark was at the home of a young businessman, an importer who also did consulting for the foreign ministry. He and his wife were very conscious of their Jewish identity and made a point of holding a number of Jewish rituals, although their overall lifestyle was very secularized. After we had finished the meal, we sat and talked while his wife took their son upstairs for his bath. We talked at length about the different prayers recited at the meal, and he touched briefly on the origins of the Sabbath in the book of Genesis. I asked whether he believed that story, whether he really thought that God had created the world in six days. He looked uncomfortable as I phrased the question, and didn't answer; I thought he was having trouble understanding my Danish, so I elaborated further, asking whether he really believed that Adam and Eve had lived in the garden of Eden and that there had been a Great Flood. Finally he looked nervously at the door, making sure that his wife was not listening in the hallway, leaned toward me, and said in a low voice, as though confessing to a crime, "Yes. Yes, I do." He then hurried to clear away the dishes, and I never brought the subject up again.

Not all Jews take this approach, of course. A substantial minority—perhaps as many as five hundred—takes the Torah very seriously and attempts to apply its prescriptions to the conduct of daily life. Such people, are referred to in common speech as "the religious" (*de religøse*); they do not constitute an organized corporate group, but both they and others in the community recognize their distinctiveness from the secular majority. Many are affiliated with Machsike Hadas, the alternative synagogue formed in 1913 (see chapter 1). "Religious" Jews grant belief a much more central role in defining Jewish identity than the secular majority does; to be Jewish, they argue, is first and foremost to have a special relationship with God, one that can be traced to God's election of the Jews as reported in the books of Moses. This relationship obliges Jews to follow an elaborate set of religious laws, making religious belief a basic ingredient in everyday practical affairs. In contrast to most Danish Jews, the "religious" tend to spend a lot of time thinking about issues of belief, and they can talk about it easily and eloquently when asked. Their specific views vary considerably from one person to another; even within families, I often met lively disagreements among husbands, wives, and children over the meaning of particular ideas. Such disagreement tends to be accepted among the "religious," even encouraged, as they indicate a real depth of thought. At the same time, though, questions about belief are treated as distinctly secondary to questions of practice. As one man put it to me, "It is of course best to have both faith and practice. If it is a choice, though, of whether to practice without believing, or to believe without practicing, it is clearly better to practice without believing." As with the secular majority, the religious regard belief as an essentially private affair, and they seldom discuss it outside the synagogue.

Indeed, in my fieldwork, I met only two people who discussed issues of belief without prompting from me: the American couple who operate Chabad House. Chabad is a Hasidic revival movement centered in Brooklyn, one that emphasizes both scholarship in the scriptures and an emotional experience of the divine. It sponsors an international network of religious centers, both to attract members to the movement and to promote religious observance among local Jews. The Chabad rabbi, Yitzchok Loewenthal, has worked tirelessly over the past several years to promote the movement, and his rhetoric is infused with the spiritualism that inspires Chabad. He describes rituals not as enactments of tradition but as connections with God, moments in which Jews can please God and come closer to an experience of Him. In Torah study sessions and Kabbalah classes, he expounds the importance of an explicit consciousness of one's connection to the divine. A number of Danish Jews told me that they enjoyed Loewenthal's approach, that this unapologetic engagement with spirituality was liberating and exciting to listen to; they did not, however, replicate it themselves. Like so many things American, it was exciting to watch but uncomfortable to perform.

This sense of faith as a private matter, as something beyond the purview of community control, makes its presence felt at an early age. One morning in 1998, I visited a Jewish nursery school in a northern suburb of Copenhagen. Run cooperatively by a group of liberal parents within the Mosaiske Troessamfund, the school served as a daycare center for children between the ages of three and six. Just before lunchtime, the teacher brought the students into a common room, where they sat on pillows and heard a Jewish story. The story for the day had to do with the creation of the world; the teacher told about the six days of creation, showing pictures of the products of each day's work. The teacher was expressive and warm, the pictures were exciting and evocative, and the children were fascinated. At the end, a little boy raised his hand and said, "Elka, is that really what happened?" The teacher, who had told the story with such ease, looked uncomfortable at the question; she responded, "Some people think so," prompting a chorus of "But is it true?" from the throng of children. She never answered directly, throwing back "What do you think?" at their repeated demands for an answer. What they thought, they would have to work out for themselves, both in childhood and later on.

RITUAL PRACTICE

Silent though they are on the issue of belief, Danish Jews can be very vocal on the subject of ritual practice. Even very secularized Jews, who make a point of disavowing any credence in Jewish myth or the supernatural, often participate energetically in some aspects of Jewish ritual, while the religiously devout build their daily lives around it. And while they leave belief to individuals, they speak freely and forcefully about their own and others' ritual practice. This pattern derives partly from the nature of Jewish religiosity; as many observers have noted, Judaism is more a religion of doing than of believing, in which ritual observance has a value independent of any feelings of reverence associated with it. It also reflects the fractious and weakly defined nature of the Jewish world in Copenhagen. Ritual makes a public statement about individual and group identity, presenting symbols and narratives that embody particular visions of the nature of the group. As factions promote their own ideas about what it is to be Jewish, public rituals become highly contested events, places to assert or challenge the dominance of particular viewpoints. The absence of a consensus on Jewish identity in Copenhagen makes almost any ritual act a part of this contest, lending significance to practices that might go largely unexamined in many places. The result is an ongoing and often angry debate about religious practice within the community, a debate that lends vitality to the community even as it divides and sometimes embitters it.

Jewish ritual practice is built on an elaborate set of laws said to have been handed down to Moses by God after the exodus of the Jews from Egypt. These laws appear in several sacred texts in the Jewish canon: the biblical books of Leviticus, Numbers, and Deuteronomy; the codification of oral laws known as the Mishnah Torah; the collection of rabbinical commentaries known as the Talmud; and the sixteenth-century codification of the law known as the Shulkhan Arukh, among others. Taken together, the corpus of Jewish law is known as halakhah. The requirements of halakhah are detailed and comprehensive, amounting to an encompassing code of conduct that touches on all areas of life. They include rules for dietary practice, holiday observance, business and legal practice, dispute resolution, and family life, among many others. Since the second century, these laws have constituted the most important unifying force among Jews in the Diaspora. Their interpretation and adaptation to local circumstances have made up the greatest part of Jewish religious scholarship for almost two millennia.

The interpretation of halakhah has always varied among Jews, from region to region and from rabbi to rabbi. The history of European Jewry includes a host of sectarian movements, each of which have proposed a distinctive understanding of how halakhah should be applied. Their understandings have ranged from rigid literalism to virtual abolition; as early as the eleventh century, for example, the Karaite movement rejected the validity of the entire Oral Torah (Ben-Sasson 1976: 441–443; Sachar 1968: 162–164). In Western Europe, these disagreements have taken on a distinctive shape during the modern period. Prior to the seventeenth century, most European Jews were not citizens of the nations in which they lived. As in Denmark, they constituted a distinct outgroup, with separate legal status and social institutions. Within their communities, local interpretations of halakhah served to regulate personal conduct and social behavior. The extension of citizenship to Jews, beginning throughout Western Europe in the eighteenth century, transformed the meaning of these halakhic rules. The laws of the state replaced the Jewish legal system, even as increasing social contacts with non-Jews challenged the cultural rules that halakhah had enjoined. Finding a new role for Jewish law in these circumstances—adapting halakhah to a Jewry that was increasingly enmeshed in a non-Jewish world—posed a common problem throughout the modern West.

After a tumultuous period of debate, three main institutional responses emerged to this problem in the nineteenth century (Blau 1966; Goldscheider and Zuckerman 1984). One movement, known as Orthodox Judaism, called for the maintenance of halakhic rules wherever possible. While incorporation into the state made many Jewish legal structures impracticable, the Orthodox movement sought to preserve the laws regarding ritual and personal conduct in what they considered a pure form. Its leaders formulated a specific understanding of the requirements of Jewish law, which became a standard for "traditional" Judaism throughout the West.[3] A diametrically opposed approach characterized the Reform movement, which argued that the Jewish incorporation into the larger culture required a thorough rethinking of the purpose of halakhah. Rather than maintaining what now seemed increasingly archaic customs, Reformists focused on defining the core spiritual and moral concepts that underlay Jewish religion and incorporating these into ritual forms that echoed those of the larger culture. As they did so, they dispensed with such cultural markers as Hebrew services, traditional clothing, and the separation of the sexes. Reform congregations came to look increasingly like churches, while Orthodox congregations remained more distinct. A third movement, Conservative Judaism, tried to strike a balance between these approaches, keeping such religious traditions as Hebrew services, but modernizing others. Each of these movements has persisted into the twenty-first century, and all claim large numbers of adherents throughout the West. Reform and Conservative Judaism dominate the United States today, claiming about 40 percent of American Jews against Orthodoxy's 6 percent (Lazerwitz et al. 1998). In Europe, by contrast, the large majority of synagogues are Orthodox; Sweden, with Scandinavia's largest Jewish population, has four Orthodox and two Conservative synagogues.

In Copenhagen, no such formal divisions have ever emerged. While activists in the congregation have sometimes pushed it in Reform or Orthodox directions, the Mosaiske Troessamfund has tried as much as possible to include all of the nation's Jews in a single institutional structure. MT literature refers to the structure as a "unity congregation" (*enhedsmenighed*), a common institution in which Jews with a variety of different viewpoints can come together. The synagogue maintains strictly Orthodox ritual forms, which its leaders describe as the only way to allow all Jews to participate. In seating, for example, Orthodoxy has strict rules about gender separation. Men and women sit separately, and men perform all of the ritual roles in the service. While Reform-oriented Jews might find an Orthodox service discriminatory or uncomfortable, nothing in their rules bars them from attending it; an Orthodox man, by contrast, could not take part in a service that required him to sit together with women. Likewise, the synagogue maintains Hebrew services, Orthodox coming-of-age rites, and strict requirements for the conversion of non-Jews, all of them essential for Orthodox participation. Its only major bow to Reform sentiments is its treatment of mixed marriages; while the MT will not solemnize such unions, it does recognize their legitimacy, and it does not ostracize or discriminate against members who contract them. These procedures allow all Jews in Denmark to engage with the MT, if not always happily, and they have enabled the MT to remain the definitive institutional voice for Judaism in the country.

This lack of formal divisions, however, does not imply any sort of consensus about religious practice among the MT's membership. Serious differences of opinion exist on the subject of ritual, differences that in many ways resemble those between orthodoxy and reform. Within the congregation, an ongoing tug-of-war pits those favoring stringent traditionalism against those demanding a modernization of congregational practice. The parties to this struggle do not constitute formally organized groups, but they do make up recognizable factions, which break down in a way similar to the division of "religious" and "non-religious" Jews noted above. The "orthodox" wing includes its most religiously active members, people who tend to define Jewishness in religious terms and who observe halakhic guidelines in their personal lives. Many of them have an affiliation with Machsike Hadas, and some attend services there. Members of the congregation's "liberal" wing are generally more secularized, and they are much less likely to observe the personal requirements of Jewish law. Most have marriages or other family connections to non-Jews, making the acceptance of mixed marriages in the community a primary concern. They tend to see Jewishness less as a matter of faith and more as a matter of cultural heritage—they often describe themselves as "cultural Jews" (kulturjøder). The liberals comprise by far the majority of the MT's membership; the orthodox, however, by virtue of their activism and certain structural advantages, command an influence far beyond their meager numbers.[4]

Many Jews fall somewhere in between these camps, and many take too little interest in congregational affairs to have a strong connection to either. But the opposition between them polarizes debate on issues of ritual practice in the MT, touching on virtually every activity in which the community engages. Planning menus in the MT nursing home, for example, requires a choice between kosher and non-kosher food; since there is only one set of kitchen facilities, the home has opted for an entirely kosher regime. Liberal residents often bristle at this imposition, which requires them to change their diets after a lifetime of standard Danish food. Likewise, an attempt to set up a café in the basement of the MT office building foundered in the late 1990s over the question of non-Jewish customers. Orthodox proponents argued against allowing non-Jews inside, in order to promote marriages within the Jewish community. Liberals, by contrast, demanded the right to bring non-Jewish spouses or sweethearts in as they saw fit. These contrasting visions of ritual obligation invest virtually any Jewish activity in Copenhagen with political tension.

Despite this tension, ritual practice among Danish Jews remains strikingly vital. As noted above, the synagogue attracts hundreds of Jews on an average Saturday, and on major holidays it is filled to bursting. Life-cycle rituals, like coming-of-age ceremonies, weddings, and circumcisions, are important family occasions with large festive gatherings. And while they do so in very different ways, a large proportion of the Danish Jews incorporate the ritual obligations of halakhah into their personal lives. They do so despite a lack of consensus on what the rituals mean, how they should proceed, or why they are important. Indeed, it may be precisely this lack of consensus which makes the rituals so enduring. Public rituals among the Danish Jews offer a variety of different avenues for individual engagement; their structure allows Jews with deeply differing notions of God and group to participate. Private ritual practice, likewise, can take a number of different forms, each implying distinctive views of the nature of Jewish identity. Together, these forms of practice create a common language through which Jews can speak to themselves and each other about what it is to be Jewish. In this very fragmented community, ritual cannot create the kinds of unity and common sentiment envisioned by the Durkheimian view of ritual function. It can, however, provide a common symbolic space, a common reservoir of symbols and meanings on which Jews can draw as they construct their individual understandings of self and community.

The best sense of this role of ritual practice comes from a direct look at the rituals themselves. We turn now to a closer look at a few of the most important elements of religious practice in Copenhagen—first the world of public ritual, and then the world of private observance.

Public Ritual

Synagogue Services

Synagogue services make up the most visible and regular public ritual in Judaism, an encounter with the sacred scriptures that occurs up to three times every day. The core of the service lies in the reading of the Torah, the five books of Moses, which must be recited in their entirety by each congregation over the course of the year. In addition, services involve the invocation of special prayers, blessings, and sacred poetry drawn from the Jewish scriptures. Orthodox synagogues hold three short prayer services, known as *shakharit, minkha,* and *ma'ariv,* every day; *shakharit* is normally held in the early morning, while *minkha* and *ma'ariv* are often combined into an evening service. On Saturday mornings, a much longer service is held, which includes lengthy readings from the Torah, a rabbinical sermon, and a variety of special prayers. Some features of these services are ordered according to specific scriptural references, such as the requirement that all involve at least ten adult men (a group known as a *minyan*). Others have emerged as traditional features over time, such as the elaborate ceremony involved in covering and uncovering the sacred scrolls.

The basic structure of these services is relatively uniform across the Jewish world, with specific Torah readings and prayers specified for each date in the Jewish calendar. The details of the ritual procedure, however, vary considerably. Even among Orthodox synagogues, specific features of services differ among different congregations. One rabbi might give sermons in the vernacular, for example, another in Hebrew; alternatively, individual congregations may leave out certain prayers or sacred poetry that they consider extraneous to the service. Reform and Conservative congregations take such changes much farther, reconsidering some of the basic features and structures of the Orthodox ritual. Both movements reject the Orthodox practice of separating men and women in the synagogue, and they allow women to take leading roles in the service. Conservatives incorporate more of the vernacular language than the Orthodox, who conduct almost the entire service in Hebrew; the Reform movement translates most of the service into the vernacular. Other variations touch on the order of ritual activities, the style of dress and head covering, and the overall length of the service. For any particular Jewish congregation, the ritual format represents a choice among a variety of options, a choice usually arrived at through a lengthy and contentious analysis of its symbolic implications.

The city of Copenhagen currently has three synagogues: the main MT synagogue, sometimes referred to as the Great Synagogue;[5] a small synagogue run by the orthodox splinter group Machsike Hadas; and a basement synagogue in the home of an orthodox leader named Erik Guttermann. The MT and Guttermann have each also set up a synagogue in the northern coastal town of Hornbæk, a popular vacation spot among Copenhagen Jews. The synagogues in "Jødebæk," as Jews sometimes call it, operate only during the summer months. Neither of the small orthodox

synagogues currently employs a rabbi, though Machisike Hadas has sometimes done so.

The Great Synagogue

The Great Synagogue lies on Krystalgade, a narrow street that bridges two
of the main arteries of Copenhagen's central pedestrian shopping district.
Across the road stands a large branch of the Copenhagen public library.
About a block down the street, at the junction with Købmagergade, Christian IV's massive Round Tower rises over the city. Thousands of shoppers
and tourists pass through the street daily, as does a steady stream of auto-
mobile traffic. Most of them pass by the synagogue without giving it a second look. A long rectangle of dirty yellow brick, the synagogue is neither
visually striking nor beautifully situated, nor is it accessible to the curious.
A tall wrought-iron fence runs the length of the synagogue property, ter-
minating in an intimidating, electrically controlled gate. Two small signs on
the gate give synagogue hours and a contact number, and a Hebrew inscription is visible over a door facing the street. Otherwise, for six days of
the week, the property gives little sign of life to the casual passerby.

On Saturday mornings, though, the synagogue comes to life. Early in
the morning, a pair of guards in windbreakers take up a position at the
gate, and a Copenhagen police car parks across the street. Soon afterward,
Jews begin to arrive for services, first just a few, then a steady stream that
continues throughout the morning. They arrive on foot; Orthodox practice forbids Jews to drive on the Sabbath, and even members who have driven in will park some distance off to give an appearance of having walked.
Most of them dress well, with suits and ties for the men and long dresses
for the women. Members of the community either nod or say "Good Sabbath" ("*gut shabbes*") to the guards, who smile back at them and return the
greeting. The guards are members of the congregation and know most of
the Danish Jews by sight. Unfamiliar faces are stopped and questioned in
English, then usually let in. In most cases, the strangers are Jewish tourists,
and the guards tell them how to get to the sanctuary. They also check to
make sure that the visitors are not carrying anything, as even a purse or a
wallet would violate halakhic rules.[6]

Once through the gate, members walk around the end of the building
into a large courtyard, half paved and half grass, with a granite Holocaust
memorial at its center. At the far end of the courtyard rises Meyers Minde,
a five-story gray brick nursing home operated by the MT. A few Jews are

usually standing in the courtyard, talking together and waving to friends as they arrive for services, but most people hurry directly into the high entrance doorway of the synagogue. There families split up—men and boys put on skullcaps (*kippot*)[7] and proceed through a hallway into the sanctuary, while women, girls, and very small boys climb the long stairway to the balcony.

After the forbidding gateway, the rather dreary courtyard, and the dark corridor at the entrance to the building, the synagogue's sanctuary is an unexpected delight. Men walk into a vast, airy room, flooded with light from the balcony windows. Gilded chandeliers and woodwork gleam against white walls, carved and gilded panels decorate the ceiling. Facing the entrance, set into the eastern wall, a broad alcove holds two tall candelabras, a podium, and, in its center, a giant curtained set of doors with a Hebrew inscription above them. Thick square columns hold up the massive balcony, and continue up to the ceiling; between them, women's faces peer downward. Near the front, centered on the eastern wall, a square railed platform (*bimah*) holds a large slanted desk. Everywhere else, row upon row of narrow dark wooden pews cover the floor. Men make their way to their personal pews, assigned and rented by the synagogue board, and pull books and ceremonial regalia from the locked cabinets in the seats.

The surprise of the synagogue derives at least in part from its contrast with standard Danish style. Danish design emphasizes geometrical planes, smooth contours, and solid primary colors; a minimalist aesthetic, embodied in Danish Modern furniture, shapes the interior environments in which most contemporary Danes live. The synagogue, by contrast, brims with lavish ornamentation. Men drape themselves in white silken prayer shawls (*tallitim*), embroidered in blue and gold, which they adjust continually throughout the service. The leaders of the service wear stately black robes and elaborate hats, with shapes that indicate their wearers' roles. The eastern wall is covered with a lattice of gilded molding, and above it golden letters emblazon a statue of the Ten Commandments. Most elaborate of all are the Torahs themselves, the massive sacred scrolls that lie behind the great doors on the eastern wall. Each is draped with an embroidered velvet cover, a silver "breastplate" on a chain, and an intricately carved pointer. When they are taken down to the *bimah* to be read, an elaborate ceremony of undressing, unrolling, and displaying them calls attention to their decoration. The atmosphere of the service involves a distinctive sensory richness, an aesthetic that differs from the world outside both in the particular symbols it uses and in the lavishness with which it displays them.

The service itself is long, intricate, and carefully choreographed, involving a number of different actors. It begins early, around 9:00, and proceeds through a long set of Torah readings, blessings, and prayers. Most of the action is centered on the *bimah*, where a professional cantor (*chazzan*) sings the texts in a rich, mellifluous Hebrew. A small chorus of men stands in front of him, by the eastern wall, singing the congregation's part in the responsive passages. The rabbi leads a few parts of the service, notably his sermon and some of the more important blessings. Most of the time, however, he oversees the proceedings from his seat in the first row, singing along with the prayers while keeping a sharp eye on the comings and goings of the congregation. A variety of other men participate as well, particularly for the readings from the Torah. Elaborate ceremonies accompany the removal of the Torah from the ark (the cabinet that holds the scrolls), the transport of the scroll to the bimah, the uncovering and unrolling of the scroll, and the beginning of each reading. At several points in the service, the scrolls are paraded through the aisles of the synagogue, where those present can receive a blessing by touching it with their prayer books or prayer shawls. Participating in any of the service's activities is an honor, and each honor has a specific symbolic connotation. Being the third person to recite the blessing before a Torah passage, for example, is a mark of particularly high prestige. The process of the ritual thus involves an unfolding social and political drama, as the rabbi calls different members of the congregation to take part in different tasks. The service goes on for a long time, usually at least four hours, and ends in the early afternoon.

As the service unfolds on the bimah, some of the men in the pews sing along. Each has a prayer book, and some of the more devout try to pray together with the cantor throughout the service. Four hours is a long time, however, and even the most diligent take breaks. They periodically stop to rest, to sit in their pews, to gaze around the room, to have a few words with friends or neighbors. The service represents, after all, the largest regular gathering of Jews in Denmark, and even the most religiously committed take the opportunity to make contact with friends and family members. This dimension of the ritual has even more attraction for the less devout, who in many cases spend almost their entire time in the synagogue socializing. Most Jews in the congregation speak little or no Hebrew; while they can sing along with a few prayers from having heard them so often, the content of the service is virtually opaque to them. They come to the service only for an hour or so, largely for the Jewish social experience it offers. They make no effort to hide this fact. Indeed, the parade of ritual movement and incantation on the *bimah* is more than matched by the motion and noise of the floor. Men wave to each other across the aisles, smiling and gesturing to newly arrived acquaintances. They move constantly about the floor, crossing from pew to pew or out to the courtyard to meet their friends. Their conversation makes a constant hubbub in the synagogue, one that swells enough periodically to drown out the cantor. When it gets too noisy, the rabbi utters an angry "Shhhh!," which produces quiet for a minute or so. The conversation stops altogether only at a few points—at key prayers like the *Sh'ma,* which everyone sings from memory, or the rabbi's sermon. Even then, their silence owes less to reverence than to politeness. During one service, I was deep in a conversation about immigration with a middle-aged businessman when he suddenly stopped talking, took me by the sleeve, and pulled me with him out of the room. I assumed that he was about to tell me something secret, and I paid special attention as he whispered to me in the hallway. After about fifteen minutes, however, he perked up, smiled, raised his voice back to normal, and said, "We can go back in now. The sermon is done."

When I describe this dimension of the service to Danish Jews, they almost always laugh. It's true, they agree, with the kind of embarrassed smile that goes with confessing an enjoyable sin. None of them, however, even the most orthodox, really seem to disapprove. This is not, after all, a Protestant service; its value does not depend on its ability to inspire faith or solemnity in the assembled throng. As long as the prayers and scriptures are recited in accordance with the law, as long as those who wish to can pray along unmolested, who is hurt by some conversation and mild disorder? And indeed, it would be wrong to view the action on the floor as an aberration, a distraction from the real business of the ritual going on up at the bimah. Socializing is a valuable part of synagogue attendance for all of the people there, and it is the primary reason why most of them attend. They join the service because it brings them together with other Jews, with family, friends, and strangers, in an environment defined by their common identity and heritage. For an hour or so—perhaps the only such hour in their week—being Jewish is not a mark of difference but of belonging. The readings and prayers matter not just because of their liturgical content but equally because of the environment they create, a setting of Hebrew song and ritual beauty within which those present can encounter each other as Jews.

This setting even extends beyond the sanctuary itself, to the gates and streets outside. The guards who monitor the entrance to the synagogue belong to the Vagtforening, a voluntary association that organizes security for all MT activities. The group's responsibility covers a wide range of events, from the daily activities of the synagogue, school, and MT offices, to special occasions held by Jewish social organizations. This work requires a lot of manpower, and indeed the Vagtforening constitutes one of the largest voluntary associations within the MT. The work also requires a high level of commitment. The group's leaders screen prospective members carefully for suitability, and they demand extensive and ongoing training for all guards. When on duty, the guards take their work very seriously. They carry concealed handguns and wear protective clothing, and they coordinate their activities with those of the Danish police. The security system makes itself felt in all MT facilities, which like the synagogue are usually surrounded by intimidating security fences and sophisticated surveillance cameras. Its human dimension, however, appears most visibly at synagogue services, where the guards stand at the border between the Jewish world within the gates and the non-Jewish world without.

It is unclear whether the MT really needs this level of security; while a Palestinian group did damage the synagogue and a Jewish business with two bombs in the mid-1980s, anti-Semitic violence in Denmark is otherwise virtually unknown. Whatever its value in instrumental terms, though, the guard service clearly offers a distinctive and appealing means of Jewish engagement for non-practicing Jews. The Vagtforening sponsors social events and holiday parties for its members, some of whom described it to me as one of their tightest social networks. At the synagogue itself, moreover, guard duty allows Jews with little interest in religious ritual to take an important role in Jewish practice. Like those inside the service, guards get to spend a few hours greeting and chatting with Jewish acquaintances. They do so, however, as guardians of the community, as armed agents protecting the safety of their friends and family within the gates. One young man likened himself to a soldier in the Israeli army, taking up arms to safeguard Judaism from a hostile world. The drama of this role can prompt a distinctive sort of affect, a cheerful bravado not commonly encountered on the tranquil streets of Copenhagen. Outside the Yom Kippur service, for example, I noticed a guard watching the gate from a doorway across the street. I had spoken with him often, on my way in and out of services, and I crossed over to say hello. I asked him if he didn't feel left out, standing there alone in the cold while everyone else went into the services. He grinned and said no, that he wasn't alone as long as he had his two friends with him.

"Which friends are these?"

"Mr. Smith and Mr. Wesson."

Few Danes, Jewish or not, ever get to make a statement like that; it smacks more of action movies or detective novels than the bourgeois world of Jewish Copenhagen. So do the radios clipped to the guards' belts, the bulletproof vests under their windbreakers, and their professional consultations with uniformed policemen. As a result, the synagogue service occasions a kind of ritual participation that appeals well beyond the ranks of the devout. It creates a setting in which Jews with little spiritual engagement can take an important leading role, and in which they can formulate a relationship to Jewish identity very different from that of the worshippers on the bimah.

The multiple faces of the synagogue service—its roles as devotional setting, as social arena, as subject of armed protection, and more—go a long way toward explaining its vitality in this largely secular society. The synagogue does not impose or demand a particular attitude toward Judaism. Rather, it creates a distinctive sort of space, a setting within which Jews with a variety of different understandings of self and community can come together and interact. This interaction does not create agreement or consensus; almost everything that occurs in the synagogue, from the ritual procedures to the seating arrangements, is the subject of ongoing disagreement and occasional angry dispute. It does, however, create a sense of community, a feeling that for all their divisions and antipathies, the people who meet in the synagogue have some common identity as Jews. The disorder of their participation—one man praying devoutly while his neighbor waves to girls in the balcony, the cantor's voice struggling to overcome the diffuse roar of an audience that ignores him—echoes the division and disorder of Copenhagen Jewry in general. The Krystalgade synagogue does not try to unify its parishioners, to create something with which all of its members can agree, but rather embraces the disunity that defines Jewish life in Copenhagen. Like many things Jewish, it produces a cacophony, but a cacophony to which every Jew can lend his voice.

The Independent Synagogues

Almost all religiously active Jews in Copenhagen belong to the Mosaiske Troessamfund, and therefore almost all of them have an affiliation with the Great Synagogue. Some of the most orthodox, however, find the services in Krystalgade unsatisfying. For all the grandeur and energy of the setting, the services lack the atmosphere of devotion and connection with the divine that these Jews look for in worship. In addition, congregational politics have set many orthodox Jews against the leadership of the MT and the synagogue. As a result, two small synagogues currently compete with Krystalgade for Copenhagen's most orthodox Jews. Their members, while part of the MT, tend to stay socially separate from the larger Jewish community and to hold adversarial views of the patrons of the Great Synagogue.

The Machsike Hadas synagogue lies in Ole Suhrs Gade, a drab backstreet crowded with old apartment buildings at the northern edge of central Copenhagen. If the Krystalgade synagogue shows few signs of life to passersby, this one shows none; except for the words "Machsike Hadas" by one of the doorbell buttons, nothing on the exterior of the building indicates the presence of a synagogue there. The building, like the others on the street, is a five-story gray brick rowhouse, with a narrow central staircase and small apartments originally built for laborers. The lower floors, however, have been converted into ritual spaces, including a surprisingly roomy and well-appointed sanctuary. This synagogue lacks the grandeur of Krystalgade, but it has an arched ceiling, a bimah and ark in rich dark woods, and cleverly designed (if rather uncomfortable) wooden pews. The building also contains a meeting room, a ritual bath (*mikva*), and rooms for a small Hebrew school (*cheder*). Machsike Hadas holds a full range of rituals there, including daily and holiday services. Its members also conduct Torah study sessions and some social activities in the meeting room. The atmosphere of the place differs radically from that of the Great Synagogue, with its sense of size and energy and lavish ornamentation. Everything about the Machsike Hadas building is small, quiet, and rather seedy, made interesting not by architectural flourishes but by a feeling of intimacy and pervasive concern with the Torah.

The synagogue on Ole Suhrs Gade has one drawback for the orthodox wing: its location, several kilometers from the Frederiksberg district where a sizable percentage of Machsike Hadas members now live. Since halakhah requires Jews to walk to services on the Sabbath, this distance poses a serious obstacle for the group's increasingly elderly population. For their benefit, orthodox leader Erik Guttermann has set up a simple synagogue in the basement of his Frederiksberg home. The room has a makeshift character, with folding chairs and photocopied wall decorations rather than pews and woodwork; still, it has everything necessary to conduct daily and weekly services. Indeed, because of its location and Guttermann's considerable prestige, it has a fairly sizable clientele, and on Friday evenings it can attract as many or more visitors than the main Machsike Hadas building.

Services at these independent synagogues have a much simpler and more serious flavor than those in Krystalgade. They lack the pageantry associated with the larger building—the elegantly attired clergy, the lavish design, the pomp and ceremony associated with the different phases of the ritual. They also lack the bustle and social interaction that permeates the Great Synagogue. The entire focus of the service is on the liturgy, which is conducted in a serious and businesslike manner. All of the men participate, taking turns reading from the scriptures or performing other ritual offices; all of them read Hebrew, and all are intimately familiar with the procedures required. Women have virtually no role in the proceedings, and they rarely attend. As with the Krystalgade synagogue, most of the service follows a preset order, with a variety of prayers and texts to be recited for each calendar day. Regular Jewish ritual does not require a rabbi, only the presence of ten or more men, and Machsike Hadas does not employ one for most services. Instead, individual men in the congregation stand up at specific points of the service and give their own thoughts on the interpretation of scriptural passages. Their opinions often give rise to disagreements, and they may produce extended and heated debates among the assembled men. Combined with the already substantial length of the service, these debates can make Sabbath services at Machsike Hadas very long affairs, for which—unlike their counterparts in Krystalgade—most of the men remain until the end.

For all its simplicity, the Machsike Hadas service produces a compelling effect, one very different from the elaborate pageant of the Krystalgade ritual. The service is not something to watch but something to do; the press of men crowded into the small rooms, working briskly through the tasks of the liturgy like soldiers pitching a camp, leaves no room for spectators. The texts and ritual action, moreover, are not a backdrop for social interaction but the heart of the event, the centerpoint around which attention and reflection turn throughout the ceremony. The result is an intense engagement with the words and ideas that the service embodies. In Krystalgade, even the most devout participants direct much of their attention to the service's social dimensions; they look around for friends as they pray, they monitor whom the rabbi has chosen for each *aliyah,* they evaluate the political messages the rabbi has embedded in his sermon. The Machsike Hadas service has a social subtext too, of course, like all rituals, but it plays a distinctly secondary role. The foremost concern is the word of the Lord, the performance of rites that members believe have been mandated to them by God Himself, and which they feel connect them with millennia of Jewish faith and history. As one orthodox man told me, "The big synagogue is interesting, and very beautiful, but when I want to feel close to God, I go to Machsike Hadas."

Unfortunately for him, doing so has become more and more difficult in recent years. The orthodox wing of the Copenhagen Jewish community is dying out, withered by a combination of emigration, intermarriage, and the secularizing thrust of contemporary Danish culture (see chapter 3). Two generations ago, Machsike Hadas could count on dozens of men to attend all of its services; today, gathering the necessary *minyan* poses a daily challenge for orthodox leaders. The group would certainly have collapsed already if not for Erik Guttermann, an elderly industrialist who has used his fortune and his considerable personal charm to keep it alive. Not only does Guttermann provide the main financial support for Machsike Hadas, he has established a number of institutions that bring observant Jews into the area: a school that gives Hebrew instruction to Russian boys, an apartment building reserved for orthodox Jews, a kosher hotel in Hornbæk, and others. When I attended his synagogue on one Friday night in October 1998, the native Danish attendance included only six or seven elderly men. However, with a contingent of visiting American students, four Russian teenagers, and Rabbi Loewenthal—all recruited by Guttermann—the room was almost full.

The incorporation of foreigners into Machsike Hadas services has one rather unexpected consequence: it has given the orthodox wing of the Jewish community a strongly international orientation. When I visited my first service in Ole Suhrs Gade, in 1997, the chairman of the synagogue asked if I would like to visit the Torah study session afterward. I said yes, and he led me into the meeting room, where an olive-skinned man with an impressive gray beard presided over six or seven young men at a table. One of the men was American, an immigrant married to a Danish Jew who had lived there for several years. I sat with him as we waited for the session to begin and asked a few questions, including some about the makeup of the Machsike Hadas congregation. I had noticed a number of people who didn't look like Danish Jews at the service and asked how normal that was. With some defensiveness, he insisted that most of the people in the group were native Danes; sure, there were a few immigrants, like him, but there were a lot more natives.

"Do the natives also do things like the Torah study," I asked, "or is that mostly an immigrant activity?"

"Of course natives do it," he replied. "A lot of immigrants do it, but so do a lot of Danes."

"But a group like this would be mostly immigrant?"

"No! There are Danes right here with us now. Some immigrants, yes, but Danes too."

At this, one of the other men at the table laughed, and said "I thought we were all Israelis!"

"Of course not!" said another, pointing to the leader. "That one's a Yemenite!"

The table broke up into laughter, and it quickly emerged that there was not a single native Dane remaining in the building. My compatriot was deeply annoyed, both with them and with me; he was working hard on blending into Danish society, and I suspect that he didn't like being grouped with other immigrants. Most native members of Machsike Hadas, however, acknowledged the presence of immigrants quite openly, and didn't seem bothered by it. Most of them have family connections in Israel, England, and the United States, and most have traveled frequently to Orthodox communities outside of Denmark. The important thing, several told me, was that you were an observant Jew, not that you lived in this or that country; when studying the Torah or making a *minyan*, national origin was irrelevant.

As a result, their ritual has a much more international atmosphere than that of the larger community. A service in Krystalgade is in many ways a parochial affair. It draws together a distinctly local community, and following its social and political nuances requires an intimate familiarity with the history and networks of the Copenhagen Jews. Newcomers often feel shut out, and even Jews who had grown up in Denmark often told me that they felt marginalized by the orientation of the MT toward "insiders." Machsike Hadas, by contrast, offers a haven to immigrant Jews, who in some cases have become much more observant since arriving in Denmark. The value placed on Hebrew gives the group a special attraction to Israelis, several of whom told me they had received a cold welcome in the larger synagogue. Their presence, along with that of visiting Americans, makes a service in a splinter synagogue an international event. In the interstices of the ritual, one often hears more English spoken than Danish, and the food and drink served often reflects Middle Eastern as well as Danish tastes. This openness to foreigners has distinct limits, of course; strangers are embraced only if they are observant orthodox Jews. Even so, the international flavor of services in Machsike Hadas belies the larger community's stereotype of the orthodox as xenophobic and reclusive.

✠ ✠ ✠

The synagogues in Krystalgade and Ole Suhrs Gade represent two opposed approaches to the dilemma of religious observance in contemporary Jewish Copenhagen. Over the past century, the Danish Jews have become increasingly diverse in their religious outlooks. Some of this diversity reflects the turbulent history of immigration and factionalization that has transformed Danish Jewry since 1900; some of it, surely, also reflects the growing individualization of religious experience in late modern culture generally. Religious groups throughout the urban West have found it increasingly difficult to maintain common rituals, as modernity has undermined the local communities within which religious observance has traditionally flourished. Whatever the reasons, establishing a consensus on religious ideas has become next to impossible among Danish Jews in recent decades. Under such circumstances, how does one create a meaningful ritual? How can Jews be brought together to worship when their understandings of what they are and what worship should do differ so radically?

The Krystalgade synagogue answers this question with a multidimensional feast, a ritual that offers a variety of avenues for individuals to engage with it. It allows members to participate in different ways and to impute different meanings to their actions. As a result, it creates a common meeting ground for various kinds of Jews, a ritual space within which different understandings of Jewish identity can interact with and sometimes confront one another. This approach has made the synagogue services among the liveliest religious observances in Denmark, and it has made them the most visible regular activities in the Jewish community. It has done so, however, at the cost of its religious message. Gorgeous as it is, the Krystalgade service is seldom theologically powerful; it does not produce the attention to texts or the debate over ideas that occurs routinely at Machsike Hadas. The prayers in the service are largely ignored, as are the rabbi's sermons and the readings from the Torah. By its receptivity to all understandings of Judaism, the service loses its ability to forcefully promote any particular viewpoint, or to seriously engage the texts at its heart.

The Machsike Hadas synagogues do promote a viewpoint, and they do engage the texts. Unlike the Great Synagogue, they do not attempt to satisfy a range of approaches to Judaism; they embrace a single viewpoint, that of the Orthodox movement, and they focus intensively on the ritual acts and theological concepts of the Orthodox liturgy. The result is an intense religious experience, a feeling of devotion and community unimaginable in the din and bustle of Krystalgade. To carry off such a ritual, however, requires a strong and unified Orthodox community. For logistical reasons, the service requires a large number of people schooled in Hebrew liturgy and devoted to ritual observance. At a deeper level, it demands a group with a common understanding of Judaism and Jewish identity, a group that can agree on what the rituals mean and how they should proceed. As the Copenhagen Jews have become increasingly diverse, such a community has become more and more difficult to maintain. Machsike Hadas has become smaller and smaller over the decades, until now it hangs on by its fingernails, dependent for its survival on a single wealthy benefactor and a shifting collection of foreigners. For all its religious force, this approach to ritual seems increasingly distant from what Danish Jews have become—the complex, decentered cacophony of selves that the Krystalgade service echoes so well.

Holiday Observance

In addition to the regular rhythm of work week and Sabbath, the Jewish calendar recognizes a number of special holidays, marking a variety of sacred, seasonal, and historical events. Some of these days focus primarily on spiritual matters. The High Holy Days of the fall, for example (*Rosh Hashanah* and *Yom Kippur*), focus on God's judgment of His people, and they call upon Jews to evaluate their lives and atone for their sins. The *Simchat Torah* festival, coming shortly afterward, celebrates the annual completion of the reading of the Torah. Other holidays commemorate major events in Jewish history, such as the Exodus from Egypt (Passover, or *Pesach*), the victory of the Maccabbee rebellion (*Hanukkah*), and the founding of the state of Israel (Israeli Independence Day, or *Yom ha-Azma'ut*). The holidays often have a strong seasonal component, and some of them involve elaborate festivities. The holiday of Purim, for example, which commemorates Haman's foiled attempt to annihilate the Persian Jews, calls for the reading of the Book of Esther in the synagogue. In most Jewish communities, the event also involves elaborate and often comical costumes, the use of shouting and noisemakers, and boisterous parties or plays afterward. During Sukkot, which celebrates the Israeli harvest, many Jews build an outdoor shelter (*sukkah*) festooned with harvest symbols in which to eat their meals. Each holiday has a distinctive character, involving specific formal observances and extensive folk traditions.

In most cases, holiday observance in Judaism extends well beyond the synagogue. Major holidays involve not only special religious ceremonies but also special home rituals, community events, and often special taboos and rules of personal conduct. On Yom Kippur, for example, observant Jews fast from the sundown that opens the holiday to the sundown that closes it; afterward, families break the fast with a large celebratory meal. During the spring holidays of Passover, a more limited taboo enjoins Jews from eating any form of leavened bread. Some rituals take place in the home, such as the lengthy ritual feast (*seder*) at Passover. Others take place outdoors; the Taschlich ceremony, for example, requires Jews to cast their sins symbolically on the waters at a natural shore. A number of holidays require Jews to abstain from work, and Yom Kippur also forbids them to bathe, have sexual intercourse, or wear leather shoes. Whatever the requirements, Jewish holidays create a total experience, a period of sacred time in which life at home, at work, and at worship take on a palpably different tone.

The particular shape of this experience varies considerably from place to place. While *halakhah* provides a basic structure to Jewish holiday observance, the application of halakhic requirements and the folk elaborations on them differ from region to region and from one theological movement to another. Some of this variation involves minor details; on Hanukkah, for example, most Eastern European Jews celebrate with potato pancakes, while Israeli Jews tend to eat doughnuts. Interaction with neighboring societies can produce more significant differences in observance. In Western Europe and the United States, for example, the importance of Christmas in the larger society has made Hanukkah the most prominent Jewish holiday despite its relatively minor theological significance. The result is a constellation of variations for each holiday, as every region, every congregation, even every family has distinctive ideas and traditions that shape its holiday observance.

The historical mobility of Jews gives this variation a special importance in community life. Very few Jews live in communities drawn from a single background; in most cases, Jewish communities include a number of different subgroups, each tracing its ancestry and its traditions to a different region. Community members also differ in their theological views and their relationships to the non-Jewish world. Such differences lead to different styles of holiday observance, some of them incompatible with one another. Any holiday, therefore, involves the juxtaposition of a number of different traditions, and to the extent that the community observes rituals together, it involves a choice among them. The holiday provides an opportunity for families and factions to highlight their own traditions and viewpoints. It also provides an opportunity for them to further their own versions at the expense of others, advancing their own dominance or challenging that of others in the political world of the community. While holidays are important occasions on which Jews come together, they are also important places for Jews to distinguish themselves from one another, and they always carry with them at least the potential for conflict.

Holidays in the Great Synagogue. Regular Saturday services in the Great Synagogue attract only a small part of the Jewish community; many more participate in holidays, however, and on major occasions like the High Holy Days, the Great Synagogue overflows with celebrants. At Machsike Hadas, enough people attend at holidays to justify flying in a rabbi from Israel to officiate. Many non-observant Jews told me that they went to the synagogue only twice a year, on Rosh Hashanah and Yom Kippur, along with their families. Most Jews also make some effort to observe the home rituals involved in major holidays. On Passover, even Jews who never attend a synagogue often take part in a seder meal, while Hanukkah serves as an analog to Jul (the Danish Christmas) for most Jewish families. Indeed, for some Jews, holidays mark the only occasion during the year when Jewish identity takes on a tangible form. Most Copenhagen Jews spend their workdays in non-Jewish offices, socialize largely in non-Jewish circles, and marry non-Jewish Danes; while Jewishness may be important to their self-identities, for some it almost never surfaces as a socially enacted experience. For them, holidays create a rare moment when Jewish identity becomes something active, a lived reality rather than a descriptive feature of the self.

Holidays have a particular resonance in Denmark because they depart so visibly from the larger cultural world. In everyday life, Copenhagen Jews tend to avoid drawing attention to their religious distinctiveness. Jewish holidays, however, have no real parallel in Danish Christian culture, and to the extent that they observe them, Jews identify themselves as religious and cultural nonconformists. The dates of Jewish holidays, for example, vary from year to year, due to inconsistencies between the Jewish calendrical system and the Gregorian one. As a result, not only must Jews excuse themselves from work to observe holidays, they must explain to confused co-workers why they do so at a different time every year. Likewise, Danish holidays have no special rules that correspond to fasting on Yom Kippur or eating matzoh during Passover. Even the Hebrew names of the holidays and the unfamiliar events they celebrate seem strange amid the pervasive Christian imagery of Denmark. Holidays therefore make Jewish identity unusually visible, highlighting the gap between the ordinary Danishness on the surface of Jewish self-presentation and the difference that lies beneath it.

Danish Jews experience this visibility as at once painful and exhilarating. Danish culture as a whole tends to value conformity, stigmatizing difference and aggressively punishing self-importance. Danes generally work hard to avoid sticking out, and Jews—who are, after all, also Danes—often find the struggle to balance this ethic against their ethnic and religious difference a very difficult task (see chapter 6). The visibility of the holidays makes this problem particularly acute. To miss work or school days, to pull matzoh out of one's lunchbox at work, to set a menorah in one's window—such actions involve a painful visibility in a culture so dedicated to fitting in. At the same time, however, they provide a legitimate escape from conformity, an excuse to be the one thing that Danes are never permitted to be: different. When they come to the synagogue for services, they can immerse themselves in an alternative identity which for most of the year remains largely covert. This release from conformity, this reveling in the otherness of Jewish identity, gives occasions like the High Holy Days an energy that sets them apart from ordinary synagogue services.

I experienced this energy myself in 1998, when I attended the Yom Kippur services in Krystalgade. Like all Jewish holidays, Yom Kippur runs from one sundown to the next, and it begins with an evening prayer service.[8] I arrived at 5:50, about ten minutes before the service was scheduled to begin. On a regular Saturday, the courtyard would have been deserted before the service, as only the most devout would have arrived early, and they would be inside; on this night, however, the courtyard was bustling, with thirty or so well-dressed Jews milling around, chatting with friends or waiting for family members. The security detail looked different as well. Not only did the guards at the gate wear suits rather than windbreakers, they also covered a larger area, with several patrolling the street outside. They all had looks of grim seriousness on their faces, conveying a palpable tension despite the absence of any visible threat to the synagogue. Near the entrance to the building I ran into an acquaintance, Asher Rosovsky, who was standing with his father by the door. We talked for a moment, and they invited me to sit with them, as the brother they were waiting for had not turned up. The invitation turned out to be fortunate—I had never seen the synagogue full before, and I hadn't realized how difficult it would have been to find a seat. We pushed through a throng into the hallway, made our way to the sanctuary, and walked down the aisle through a sea of suits and white prayer shawls. Our progress led us through a cross-section of the congregation's history. Reserved seats are passed down in the synagogue, from father to son; the best seats, those nearer the Eastern Wall, belong to the families who have maintained the longest continuous presence in the congregation. We sat a few rows from the back, in the center of the floor, a location that reflected the Rosovskys' status as an old family in the Jewish community. A few rows behind us sat the largest part of the congregation, descendents of Jews who had immigrated from Eastern Europe at the turn of the century. Behind them sat members of the wave of Polish immigrants from 1969, and in the last rows and standing in the aisle behind them came the recent immigrants from the Middle East. A similar structure obtained among the women in the overflowing balcony. The easy informality of the Saturday services, with people arriving and departing at random, praying and socializing at intervals amid the placid backdrop of the ritual, had given way here to a pageant, at which punctuality and self-presentation mattered acutely and in which the congregation's hierarchies of lineage and prestige were vividly and unmistakably rendered.

The Yom Kippur holiday marks the most momentous day of the Jewish ritual calendar and one of its most solemn; it focuses on the sins that Jews have committed against God, and its prayers enumerate mortal failings and ask for divine forgiveness. While by its conclusion it often produces a cathartic joy, its subject is not a cheerful one, and most Kol Nidre services I had attended previously had had a contemplative and almost mournful tone. At the Great Synagogue, however, the atmosphere was unmistakably festive. As the ritual prepared to start, the men around me seemed excited. They looked around for friends, smiling and waving when they saw them, and fiddled absently with their white prayer shawls and white *kippot*. When the service finally began, they sang along ardently with the prayers that they knew. While few of the men present understood Hebrew, they had been hearing these songs annually for most of their lives, and most of them could sing a number of the early prayers from memory. When the service reached particularly well-known passages, such as the *Avinu Malkeinu,* the effect was electric. The thousands of voices singing in unison, the sea of white garments, the haunting rise and fall of the old melodies, produced a feeling of grandeur and catharsis that is difficult to describe. I saw little sign of concern with the solemn message of the prayers; while the liturgy focused on atonement, the participants seemed focused on community, on the remarkable experience of being together with so many Jews in so magnificent a setting. I saw a number of men I had interviewed in secular settings, men who had declared a complete disinterest in religion and admitted plainly that they knew little or nothing of Jewish liturgy, now decked out proudly in ritual garments and singing along loudly with the congregation.

The general interest in the prayers and formal ritual did not preclude other meanings for the service. Indeed, after an initial focus on the liturgy, most people in the crowd turned, as on Saturdays, toward the social side of the event. They followed loosely along in their prayer books, tuning in for key passages and joining the chorus for well-known prayers, but they directed most of their attention to connecting with friends and family members. Doing so was more difficult than usual; the crowd made it hard to see more than a short distance, and it was impossible to move from pew to pew for visits. Still, the press of people meant that everyone had some acquaintances within reach, and the structure of the seating meant that most people were sitting with relatives. A buzz of conversation soon rose into a din, and periodically it became impossible to hear the cantor. An angry "Shhh!" periodically dimmed the noise, but it would quickly pick up again. At one point, with the cantor looking downright angry, the rabbi briefly stopped the service. He stood up next to the bimah, a smile on his face, and made a theological observation about the significance of the passage that had just concluded. Conversation stopped completely while he spoke, and when he sat down there followed a few minutes of real quiet. Within ten minutes, however, the noise had resumed, and it continued until the end of the service.

Services continued throughout the following day, concluding only as the holiday ended at sunset. Attendance was lighter than in the evening but still quite heavy, and I experienced the other side of the seating system; sitting with the son of a 1946 immigrant, I found myself stuffed into an overcrowded pew under the balcony in the back corner. The service had a much less festive feeling than the evening before, and many of those in attendance looked fairly somber. Their mood did not reflect the solemnity of the service so much as hunger. *Halakhah* requires Jews to fast on Yom Kippur, and those who had done so were feeling the effects strongly by late morning. They particularly felt the absence of coffee, which Danes drink in great quantity; my seatmate was in a foul mood from caffeine withdrawal and had already had a spat with his wife on the ride in. As the day wore on, however, spirits lifted, and the service took on the air of a family reunion. The courtyard in particular became a lively gathering place. Friends would spot one another across the pews, motion each other toward the door, and adjourn to the courtyard to talk. Since children do not fast, the area became a picnic ground in the afternoon, alive with children running across the grass and dodging between the adults as they played. Most people came and went several times during the day, stopping in to services for a while, socializing in the courtyard for a while, then leaving the synagogue grounds for an hour or two, sometimes making a stop at a restaurant along the way.

As always, the modes of participation in the service varied, as Jews of differing theological persuasions took different approaches to their engagement with the ritual. Toward the end of the day, for example, the synagogue was once more full as the service neared its end. I had spent most of the day talking with people, and I had worn out my voice; since I was attending a dinner afterward with the Moskowitz family, I had decided to sit quietly for the rest of the service and try to get it back. I found an aisle seat halfway back in the sanctuary and tried to immerse myself in the sound of the prayers. The cantor's voice was really quite beautiful, and many people in the congregation were singing along with him. I found myself looking forward to the climactic event of the ritual: the blowing of the shofar, a trumpet made of a ram's horn, which signified the end of the High Holy Days and the closing of God's book of life for another year. As I sat with my eyes closed, listening to the service, I suddenly felt a rough tap on my shoulder. Looking up, I saw Harry Moskowitz standing over me, his coat buttoned and an impatient look on his face. "Come on," he said, "we've got to get out of here if we're going to beat the traffic." I stood up, a little disoriented, and followed him out into the hallway, where I could see his parents and siblings already headed out the door. As I put my coat on, one of his older relatives appeared behind us, visibly agitated. "You're not going?" said the old man. "Don't you want to stay and hear the shofar?" Harry hustled me out and called back over his shoulder, "No time, Uncle Lev. Why don't you record it for us and we'll play it back later?" The service clearly had a different meaning for the two men—a holy occasion of atonement and closeness to God for one, a social occasion for the other, an occasion that, for all its importance, did not justify sitting in traffic or arriving late for dinner.

Decisions like this one—how long to stay, whether to fast, whether to drive—present a basic problem for Jews at Yom Kippur, and indeed at Jewish holidays generally. Holidays ordinarily involve the alteration of everyday routines, whether through fasting, through abstaining from work, through eating special diets, or through special home activities. Jews must decide whether and to what extent they will abide by these prescriptions for their private behavior. This question applies even to those who do not attend services—in their work, their diet, their movement, and so on, they must either accede to or depart from a detailed model of "correct" Jewish action. The day before this service, for example, I had interviewed Dagmar Cohn, a native Copenhagen Jew married to a non-Jewish businessman. I had asked her whether she would attend the service, and she said probably not; she never attended the Great Synagogue, which she regarded as offensively male-biased. She preferred, she said, to observe the holiday herself, by staying home from work, fasting, and spending time thinking about her spiritual life. On this occasion, however, she would not do even that, since an important project was under way at her office, and she felt obliged to contribute to it. She would walk to work instead of taking a bus, and she would eat only enough to allow her to do her work properly, but her obligations to her co-workers came before her obligations as a Jew. For Dagmar, as for all Jews, days like Yom Kippur force a conscious evaluation of the salience of Jewishness to her larger identity. Whether or not she observes the holiday's requirements, she has to think about them and make a decision about what she will do. The very intimacy of the holidays, their concern with the details of personal conduct, make them powerful moments in the construction of the Jewish self.

I had a taste of this experience myself as I sat in the synagogue during the early afternoon service. In an attempt at good participant observation, I had fasted since sundown the night before. Along with my hosts, I had skipped breakfast, feeling the grumbling of my stomach as a token of virtue as I had watched the children eat their oatmeal. For much of the morning I had felt fairly pleased with myself. The privations of fieldwork in Copenhagen are normally very mild, and suffering a little hunger for my research made me feel more like a real anthropologist. By the early afternoon, however, the romance began to wear a bit thin. Hunger always puts me in a bad mood, and my view of my surroundings had begun to turn sour. I began to reflect that I had sat through four hours of prayer in a language I didn't understand, in what I had always considered the most uncomfortable pews in the western world; I found myself growing resentful of the service and annoyed with the people around me. It occurred to me that I could hardly do interviews in this mood, and that if things kept on this way I would not be fit company for the Moskowitzes that evening. Perhaps, I thought, I had a professional responsibility to slip out for a snack. Perhaps the value of full participation in the ritual wasn't as important as the value of full alertness and good humor. The hungrier I became, the more compelling the arguments seemed that pointed toward my getting some lunch.

I held out for a while, but around 2:30 hunger managed to conquer virtue, and I slipped guiltily out of the sanctuary. I threaded my way through the crowd in the courtyard and attached myself to a group that was going out the gate. Once on the street, I hurried out of Krystalgade and down Købmagergade, aiming for the hot dog stand at Amagertorv. It lay about half a mile from the synagogue, far enough that I wouldn't be seen by people on their way to services. Still, I felt conspicuous in my dark suit amidst the tourists and shoppers, and I fervently hoped that none of the Jews I knew would spot me. There was a small line at the stand, in which I tried to look inconspicuous before placing an order and paying as quickly as I could. As I sat down on a nearby bench, it occurred to me that I had broken a startling number of rules in a very short time. I had used money on Yom Kippur, a serious infraction; I had ordered food, a violation of the fast; the food in question was a pork hot dog, perhaps the most direct possible violation of kosher rules; and I was washing it down with a bottle of chocolate milk.[9] I wondered what the Jews would think if they saw me then, and I looked quickly around to make sure there were none there. To my horror, I immediately saw two of them, middle-aged men I had seen in the synagogue, seated at a table across the way. In the other direction, I saw a group of three young men in dark suits, standing in the shade of the Illum department store. Women would be harder to spot, I thought, since they don't wear suits, but surely there must be some in the crowd as well, and I began to panic at what they all would surely think of me. Immediately afterward, however, came the realization that the middle-aged men were seated in front of a shwarma restaurant, not on a public bench; and as I looked again at the young men, I saw that two of them held hot dogs. One of them looked up and returned my gaze. I waved at him tentatively with the remaining half of my hot dog, and he smiled back. The middle-aged men got up from their table, left some money, and ambled slowly in the direction of the synagogue. I sat back, relaxed, and took a long drink of chocolate milk, feeling then as close as I ever have to being fully part of the world of the Copenhagen Jews.[10]

Life-Cycle Rituals

Holidays and Sabbaths move the Jewish community through the year, structuring the flow of time and marking the key passages in its cyclical course. Ritual does the same thing for individual Jews. Like all religions, Judaism conducts a series of rites to mark major transitions in the life cycle. The major ones include the circumcision of baby boys, called a *brit milah,* carried out at seven days of age; the coming-of-age ceremony, known as the *bar mitzvah* for boys and the *bat mitzvah* for girls;[11] the marriage ceremony; the conversion of non-Jews to Judaism; and the burial service. Except for conversions, each of these rituals involves both a formal liturgical procedure and an accompanying family celebration. As with other Jewish rituals, the practice of these rites varies among regions and theological movements, and it changes form over time. Circumcision, for example, has periodically fallen out of favor in Jewish reform movements, and many liberal Jews in recent decades have developed rites for girls that echo the Orthodox rites for boys. In some cases, therefore, life-cycle rituals have been a key focus of disagreement among Jews, sites in which different regional and cultural traditions come into direct and visible conflict.

As they did with holiday and Sabbath rituals, the Jews I interviewed in Copenhagen expressed a range of views on the importance and proper form of life-cycle rituals. Attitudes toward bar mitzvahs, for example, differed strongly between the orthodox wing and the more liberal majority. While the orthodox tended to take the ceremony's religious dimension very seriously, most Jews I interviewed saw it as a formality, one important principally because it involved a lot of gifts. Relatively few liberal Jews expressed much interest in the marriage ritual, perhaps because so few of them had experienced it; since the MT does not solemnize mixed marriages, the great majority of Copenhagen's Jews opt for civil ceremonies. The ritual that seemed to have the strongest and most uniform support was circumcision. On that subject, all of my interview subjects were emphatic: it had to be done. Even very secularized, intermarried Jews insisted firmly on circumcision for their sons, despite the often horrified reactions of their non-Jewish spouses. Their justifications for the practice varied. Several men told me, with a smile, that it was important for a boy to look like his father. Others stressed its connection to the pact between God and Abraham. Most could not articulate any reason why they took this ritual so much more seriously than, say, a wedding, but they supported it nonetheless. One woman pointed out that the MT evidently had the same outlook, since circumcision is the only ritual service that is provided for members without any fee.[12]

The tensions that surround life-cycle rituals in Judaism take on a special form in Copenhagen, due in large part to two of their features. One is the sharp contrast between key elements of Orthodox ritual practice and the cultural patterns of Danish society. Orthodox Jewish rituals, for example, differentiate strongly between males and females. The male circumcision ceremony has no female analog; likewise, since women cannot participate in an Orthodox service, *bat mitzvah* ceremonies lack the reading from the Torah that forms the heart of the *bar mitzvah*. Conversely, since Jewish descent is reckoned through the mother, a Jewish woman married to a non-Jewish Dane may have her son circumcised, while a Jewish man married to a non-Jewish woman may not. Such patterns clash with the aggressive gender egalitarianism of contemporary Danish culture, and they can make Jewish ritual seem flagrantly discriminatory and unfair. Similarly, the authority of the rabbi to control these ceremonies can grate on Danish sensibilities, which regard services like burial and marriage as basic rights for all citizens. As individuals go through these rituals, therefore, they often find themselves torn between Jewish and Danish viewpoints, forced to confront in stark terms the oppositions between elements of their own identities.

The other distinctive feature of life-cycle rituals is the strong influence of the chief rabbi on individual participation. Unlike the Sabbath or holiday services, Jews cannot decide for themselves to take part in a wedding or a brit milah; these events require the approval and the active participation of the Jewish Community's rabbi. In some cases, like conversion or coming of age, they involve extensive rabbinical instruction. As a result, the rabbi can set conditions for performance of the rituals, using his own judgment as to how they should proceed and who should be eligible for them. This authority is most visible in regard to conversion. Some candidates for conversion receive approval in a matter of months, while others must wait for years. In all rites of passage, moreover, the rabbi dictates the format and style of the event. As a result, rites of transition have little of the flexibility that characterizes so much of Jewish ritual in Copenhagen. A wedding or a circumcision does not offer a range of styles of participation but a fixed format enforceable by the man who symbolizes the authority of the congregation.

These two features—their contrast with Danish norms, and the centrality of rabbinical authority—make life-cycle rituals powerful moments in the struggle to define and control the meaning of Jewishness in contemporary Copenhagen. As individuals encounter these rituals, they find themselves forced to confront issues of self-identity that can ordinarily be elided in everyday life. A man may easily combine Jewishness and Danishness most of the time, tailoring his participation in Jewish affairs to integrate with his life in a largely non-Jewish world; when he faces the decision of whether to circumcise his son, however, he must address the ambiguity of his identity directly and make a clear choice in relation to it. At the same time, these rituals express official Jewish Community positions on the nature of Jewish identity. The rabbi's decisions are visible statements of Community policy, and Jews from all factions try to influence them. As Jews seek to define their own identities, and as they seek to defend or challenge the meaning of Jewishness endorsed by the MT, they inevitably focus their attention and emotion on rites of passage.

This focus makes life-cycle rituals some of the most intense sites of conflict between different factions within the community. The orthodox and liberal wings of the congregation, for example, have an ongoing disagreement over the place of non-Jewish spouses in MT affairs. The orthodox tend to view intermarriage as a serious threat to the Jewish community, and they try to discourage the practice wherever possible. They oppose any formal role for non-Jewish spouses in the community, expressing doubt about the sincerity of non-Jewish spouses who wish to convert. Liberals, by contrast, generally see intermarriage as a fact of life, and they argue that reaching out to non-Jewish spouses serves the community better than excluding them. These views are often intensely felt, since they bear directly on individual experience. Most of the orthodox have made special efforts and sacrifices to contract a Jewish marriage, while most of the liberals have either a spouse or a parent who has been affected by exclusionary policies. In concrete terms, this dispute finds expression most strongly in discussions of transition rituals—conversions, bar mitzvahs, and especially funeral rituals. The MT policy of excluding non-Jewish spouses from the Jewish cemetery, for example, aroused fury among some of my informants, who anticipated being unable to bury their parents together (see chapter 4). Feelings about the relative importance of Jewish identity and family bonds, normally unspoken, come vividly to the fore in the context of life-cycle rituals.

The intensity of such feelings makes transition rituals a political minefield for the chief rabbi. As he makes decisions about ritual format and access, he almost invariably offends one or another faction in the community, and the hard feelings do not dissipate quickly. In some cases, decisions have threatened the tenure of rabbis. The dismissal of Tobias Lewenstein in 1913 stemmed from objections to his stringent conversion requirements (see chapter 1). Seventy years later, Bent Melchior's lenient policies on conversion created strong dissatisfaction among the orthodox, some of whom nearly managed to unseat him in 1982. When Bent Lexner succeeded Melchior in 1996, he brought a reputation for conservatism that unsettled many liberals; in my interviews, their fears centered largely on what he would do about conversion, burial, and coming-of-age ceremonies.

At the same time, life-cycle rituals offer rabbis one of their most potent political resources, a mechanism through which to cultivate particular constituencies within the congregation. For all the resentment it created among the orthodox, Bent Melchior's approach to conversion won him many friends in the liberal wing of the congregation. He took a relaxed approach to other transition rituals as well, giving him avid admirers even among those who disagreed with his theological views. One man I interviewed, for example, decided at age eleven that he would not take instruction for the bar mitzvah. He had come to the conclusion that he did not believe in God, and he felt that going through the ritual would be hypocritical. His parents, scandalized, sent him to Melchior, who sat him down and listened to his argument. When he had laid out his position, Melchoir nodded and agreed with him: "Feeling as you do," Melchoir said, "it would be wrong for you to take a bar mitzvah." At the same time, however, Melchior told him to keep in touch, to be active in the Jewish community; if at some future point he changed his mind, he would be welcomed to instruction then with no hard feelings. Even now, twenty-five years later, the man told me, he was still impressed with Melchior's style. "Had he told me that I was being a bad Jew, had he insisted I go through the ritual, I would have walked out and never gone to the synagogue again. This way, he made me feel like he was taking me seriously, and I've been a supporter of his ever since." Such supporters saw Melchior through a variety of conflicts with the MT board,. Deftly handled, life-cycle rituals can provide political capital as easily as political pitfalls.

The coming-of-age ceremony for girls illustrates the possibilities. As noted above, Orthodox practice distinguishes strongly between male and female roles in worship, with women largely forbidden to take an active part in synagogue services. This division has a direct bearing on the bar mitzvah, in which the initiate sings one of the week's Torah passages. Since girls cannot take such a role in the service, Orthodox Judaism has generally had no rite for them which corresponds to the *bar mitzvah*. Liberal Jewish movements have objected to the implicit sexism of this division, and they have established *bat mitzvah* ceremonies that mirror the *bar mitzvah*. In Copenhagen, this situation creates a quandary. The Orthodox orientation of the synagogue makes a true *bat mitzvah* impossible, and any attempt to institute one by the rabbi would provoke outrage from the orthodox wing. At the same time, the great majority of the congregation supports Danish attitudes toward gender egalitarianism, and the lack of a *bat mitzvah* would alienate many young Jewish women and their families. On its surface, this situation would seem hopeless for the chief rabbi, who stands to offend an important constituency whichever way he turns. By managing it carefully, however, the MT has worked out a system over the last decade that has made *bat mitzvahs* among the most popular, positive, and universally praised rituals in the congregation's life.

The key to this success has been the avoidance of a direct parallel between the bar and bat mitzvahs. The girls' ceremony does not involve reading from the Torah, and it does not take place during the synagogue service; it takes place, rather, immediately after the synagogue service, while the congregation is still gathered in the sanctuary. With the rabbi standing in the pulpit, the girl walks to the bimah, faces the congregation, and reads an essay that she has prepared over the last several months. The subject, chosen in consultation with the rabbi, usually touches on the roles and responsibilities of Jewish women. In the *bat mitzvah* that I watched, the girl discussed the importance of creativity to a Jewish woman's life, using as her model a Hungarian Jewish poet who had survived the Holocaust. After the reading, which may take five to ten minutes, the chorus sings a hymn, and then the girl turns toward the rabbi, who makes a speech welcoming her to womanhood in the community. The group then adjourns to a reception in the courtyard, after which the girl's family holds a celebration in their home.

This ceremony neatly addresses the concerns of both orthodox and liberal factions within the community. Falling outside the formal synagogue service, the *bat mitzvah* does not violate any halakhic rules, even though for practical purposes it is part of the same event. And while girls are not allowed to read from the Torah, they are allowed to do something much more appealing to liberal families: they are allowed to speak their minds, to stand in the center of a normally masculine space and declaim their views of Judaism and gender. The ceremony embraces the kinds of modern Danish values so dear to the hearts of non-orthodox Jews, allowing for individualism, for creativity, for self-expression, for the rethinking of traditional worldviews. Indeed, most of the people I spoke to about the *bat mitzvah* preferred it enormously to the *bar mitzvah* and felt that the girls had got the better end of the bargain. Boys, after all, merely fit themselves into a part of the service, having memorized an obscure passage of Hebrew that they would forget within weeks. Girls got to say something original, something that they thought was important and memorable. As they did it, moreover, they received the riveted attention of the congregation, much more so than the boys at their reading or even the rabbi during his sermon. As a result, the ceremony has managed to accommodate both ends of the congregational spectrum, both those who want to maintain the integrity of traditional gender rules and those who want to validate the equality of women. In doing so, it has become one of the few things about Jewish ritual that most people in the MT can agree on.

Private Religious Observance

The public rituals of Judaism do not begin to exhaust the religious observances mandated by halakhah. Beyond these formal and collective activities, the Torah and Mishnah lay out an extensive set of rules and commandments for the everyday behavior of individual Jews. These prescriptions span a broad range of personal behavior, from methods of individual prayer and worship to details of dress, diet, and family life.[13] If observed rigorously, they can structure almost every facet of personal life, infusing the most mundane and private daily activities with a religious meaning and a distinctive cultural tenor. Many scholars have ascribed the endurance of Jewish culture in the Diaspora in large part to their unifying power. A description of Jewish religiosity, therefore, must include not only the formal and visible ritual activities of the synagogue but also the extent to which religious law permeates the private action of individuals.

Many of the rules for personal behavior are quite specific, imposing such requirements as the separation of milk and meat or the duty of visiting the sick. Nonetheless, as with public ritual, applying these laws to actual action involves a considerable amount of interpretation. Even among Jews who strive to abide by all of the traditional commandments (commonly numbered at 613), innumerable ways exist to apply them to the changing circumstances of everyday life. Such variation is easily visible in places like New York, where arcane issues of interpretation have sparked rivalries and visible cultural distinctions among ultra-Orthodox Hasidic movements. The variety extends much further for the majority of western Jews, who have no interest in following traditional law as a whole. Most Jews in Western Europe and the United States pick and choose among the commandments, adopting a few that they find meaningful and ignoring the rest. As they implement these laws, moreover, they generally interpret and modify them in ways that fit in with their larger lives and with the national cultures that are as much a part of their identities as their Jewish backgrounds. Patterns of private observance, therefore, vary not merely from congregation to congregation but from household to household and person to person within any Jewish community.

In Copenhagen, this variation makes private observance a particularly sensitive medium for negotiating the differing understandings of Jewishness that run through the community. As noted earlier, the relationship between Jewish and Danish identity poses an ongoing problem for the Copenhagen Jews, both as individuals and as a community. As individuals go through life, they must repeatedly formulate and reformulate their understandings of how these conflicting elements of their identities relate to one another. This task poses a problem not only for individuals but also for families and groups, which must integrate these different views into collective activities and representations. Personal religious observance provides a language through which to do this. By following or not following the rules, and by interpreting the ones they follow in specific ways, Danish Jews let both themselves and each other know who they think they are; as they find ways of integrating their different styles of observance, Danish Jews merge those different selves into a group.

Seen from an Orthodox viewpoint, most Danish Jews ignore almost all of the traditional commandments. The great majority of Jewish families do not hold formal Sabbath meals or other home rituals; the great majority of men never wear a Jewish headcovering outside of the synagogue; the great majority of Jewish women seldom if ever enter a ritual bath (*mikvah*). The distinctive hairstyles or modes of dress that some Orthodox groups associate with strict observance are all but unknown in Denmark. One could argue that halakhic rules exercise a more covert influence on personal behavior; one of my informants, for example, suggested that Jewish laws about charity and ethics lend a particular character to Jewish political participation in Denmark. On a conscious level, however, most Jews associate religious law with specific ritual obligations, and they concern themselves with only a very few of these. Two of these requirements stand out: the obligation to rest on the Sabbath, and the set of laws regarding dietary practice. Not all Jews follow these rules; indeed, for Sabbath observance, most of them do not. But in both of these areas, most Jews are aware of the relevant regulations, most make a relatively conscious decision about whether to abide by them or not, and all activities undertaken by Jewish groups must take them into account. As a result, Jews tend to use them as a shorthand for assessing their own religious observance. Asked whether they considered themselves religious, many Jews I spoke with gave answers like, "I don't keep kosher, if that's what you mean" or "Well, I wouldn't hesitate to take a train on the Sabbath." The general awareness of these rules and the many approaches that individuals can take to them make them particularly useful symbolic media. We can get a sense of this role with a brief look at kosher observance.

Dietary Practice in Jewish Copenhagen.[14] Kosher observance involves an elaborate system of practices and taboos related to food, derived from a set of scriptural commandments traditionally said to have been given to Moses by God (Novak 1987; Klein 1979: 301–378; see also Douglas 1972: 41–57). These commandments forbid the eating of certain animals, most notably pig; they also prescribe distinct methods for the slaughter and preparation of animals that are not forbidden. The commandments have been augmented by a code of practice, known as *kashruth,* intended to ensure that they are never violated. These codes regulate domestic arrangements in great detail. A commandment that a kid should not be seethed in its mother's milk, for example, is extended to a complete separation of milk and meat in the kosher home. Not only should one avoid serving milk and meat at the same meal, one must serve them on separate sets of dishes, and these dishes must never come in contact with one another or with the same cleaning instruments (Klein 1979: 359–378). Such detailed prescriptions require an extensive infrastructure, both inside and outside the home. At the very least, kosher housekeeping demands a kosher butcher with an approved ritual slaughterer, as well as separate milk and meat facilities in the kitchen; ritual observance also calls for kosher wine, special breads, and a variety of other specialized products. In the Jewish communities of premodern Europe, these requirements were one of the key mechanisms through which group cohesion and isolation were maintained (Zborowski and Herzog 1952), and they continue to perform these functions in some Jewish groups today (e.g., Sered 1988; Dahbany-Miraglia 1988).

Like all ritual requirements, kashruth allows abundant room for differences of interpretation and practice. Some differences involve the application of kashruth to foods like turkey or buffalo meat, which are not mentioned in the Bible and do not fit neatly into kosher categories. Others involve the stringency with which kosher rules should be applied. Some Jews, for example, merely avoid meats traditionally regarded as non-kosher (*trayf*). Other Jews limit themselves to kosher meat, derived from animals killed by a ritual slaughterer in the scripturally approved way. Still others maintain a higher standard, *glatt* kosher, which requires that the animals' internal organs be certified free of taint. In general, while most Jews agree on the basic principles of kosher practice, dietary law has involved the same sorts of variation from place to place and viewpoint to viewpoint as other Jewish ritual requirements.

In Copenhagen, this variation occurs throughout the Jewish community. Jews range widely in their approaches to kosher practice, from strict adherence to unabashed rejection, with many options open in between. Most Jews depart from strict observance to some extent; relatively few maintain divided kitchen facilities, for example, and not many buy kosher meat exclusively. No statistical data on dietary practice exists, but kosher observance has clearly declined in recent decades, to the point at which the basic infrastructure necessary for its practice has come under strain. In 1920 the city had about a dozen kosher delicatessens; today two remain,[15] neither of which could exist without external subsidies.[16] Kosher practice does not, moreover, have the powerful moral force in Denmark that it does in many Jewish communities. In interviews, many of my informants cheerfully acknowledged eating pork now and then. They were not embarrassed by the statement, nor did they regard it as at odds with their Jewish identity. Once, while at a restaurant with several Jews, I mentioned that I had never had a pork dish called *flæskesteg;* one of the party promptly ordered some and showed me how to eat it. Conversely, some of my informants cited their use of kosher meat as an example of their strong commitment to Jewish living. In many Jewish communities, kashruth is a basic template on which cuisine is built, an assumed precondition of all cooking and eating. In Copenhagen, by contrast, keeping kosher is a deliberate act, one that individuals undertake at the times and places when and where they wish to emphasize a Jewish identity.

That statement comes at a cost. Different levels of strictness in kosher observance imply different amounts of investment, both of time and of money. Finding even ordinary kosher meat can pose a challenge in Denmark; only two shops in the country sell it, and only one sells it fresh. Shopping at these stores involves considerable inconvenience, especially for Jews living in the suburbs. It also involves considerable expense. A piece of ordinary kosher beef costs up to 50 percent more than an equivalent piece of non-kosher beef; costs for *glatt* kosher meat, which must be imported, run even higher. Moreover, kosher shops offer mainly high-quality meat, making it difficult to economize with cheap cuts. Beyond the costs of the food itself, kosher meal preparation can involve substantial outlays. Separating milk and meat involves at least two sets of dishes, cookware, utensils, cabinets, and washing materials. Further refinements, such as separate dishwashers, counters, sinks, and sometimes even kitchens, can raise the cost to thousands of dollars. For all but the very wealthy, strict kosher observance implies economic tradeoffs, and kosher families are keenly aware of the costs involved.

Keeping kosher, then, makes an expensive statement about the importance of Jewish distinctiveness. Only a deep commitment to Jewishness as a separate cultural identity can justify the costs that stringent observance incurs. For most Copenhagen Jews, who regard themselves as Danish as much as Jewish, such a statement makes little sense on a daily basis. They buy kosher meat occasionally, perhaps at major holidays or bar mitzvah parties, but not for everyday consumption. Many of them do avoid pork when possible, but circumstances sometimes make that difficult; many will eat pork at a non-Jewish home or festive dinner, for example, rather than offend a host by refusing it. Jews who do keep kosher tend to fall into two groups. Some belong to the congregation's orthodox wing, which insists on the strict observation of Jewish law in all contexts. Others use kashruth as a means of expressing a strong identification with Jewish culture in the absence of religious faith. Kosher observance allows "cultural Jews" to assert the distinctiveness of the Jewish heritage without endorsing any particular theological view. As they decide what foods to eat for themselves and to serve to their guests, then, individual Jews manifest their understandings of the meaning and importance of Jewish difference in Denmark.

The mechanisms of kosher observance also allow it to signify various group affiliations within the Jewish community. Through the specific kinds of kosher meat they buy and the merchants they patronize, individuals can affiliate themselves with the different factions and subgroups that make up Jewish Copenhagen. The two kosher shops, for example, are operated and patronized by members of distinct Jewish factions. The smaller of the two, Samson's Kosher, lies near the Ole Suhrs Gade synagogue, and it is operated by members of Machsike Hadas. The other shop, the Kosher Delicatessen, lies outside the center city, near the affluent suburbs where many of the more liberal Jews live. Both businesses rely on connections with Jewish subgroups for their clientele. The Kosher Delicatessen has strong ties to the formal Jewish Community, which supports it to maintain a supply of fresh kosher meat. The Jewish Community has worked actively to keep it open during slow periods, recruiting the current manager and encouraging members to patronize it. Bent Lexner serves as the ritual slaughterer, and the business carries his personal seal of approval. To buy one's meat at the Kosher Delicatessen implies a gesture of support for the larger Jewish Community, a contribution to the kind of Judaism represented by the Krystalgade synagogue. Patronage of Samson's, by contrast, connotes an affiliation with the more orthodox subgroups. Members of Machsike Hadas buy from Samson's as much as possible; while only the Kosher Delicatessen offers fresh meat, the most orthodox often buy frozen meat in order to keep Samson's going. Samson's also sells the only *glatt* kosher meat in Denmark. For much of the twentieth century, *glatt* was associated primarily with the lower-class Eastern European Jews, and many older "Viking" families regarded it with some scorn. Samson's also caters to the few Jews with special kosher requirements related to ultra-Orthodoxy or Hasidism. The Loewenthals, for example, need some items that conform to the standards of the Lubavitch movement, and I was told that they had arranged to import it through Samson's. Where shopping at the Kosher Delicatessen marks an affiliation with mainstream Jewish observance, then, shopping at Samson's indicates an attachment to the orthodox or immigrant fringe.

Within the context of Jewish social life, then, the level and style of kosher observance imply different attitudes toward both religious belief and group identity. These implications come into the open particularly in the context of entertaining. When hosting a meal, Jews must establish the level of kosher observance required by each guest and devise a menu that accommodates all parties. This process manifests each participant's understanding of and affiliations within the Jewish world. The meals that emerge serve as a palpable symbol of the group gathered together, a physical embodiment of the differences and accommodations implicit in Jewish social life. The consumption of the meal—or the refusal to consume it—constitutes an implicit acceptance or rejection of those differences. The complexity of kosher symbolism makes every group meal a symbolic event, a place at which to negotiate and express the nature of Jewish community.

This process becomes particularly important in Jewish homes, as members of Jewish families try to integrate their varying conceptions of Jewish identity. Jews in Denmark have a hard time finding Jewish spouses, and a majority do not; it is even more difficult to find a Jewish spouse from one's own social faction and with the same religious views. Accordingly, almost all Jewish marriages are mixed marriages, in that they involve significant differences in the religious views and backgrounds of the partners. These views must, to some degree, be harmonized; to become a family, rather than just a collection of individuals, the members of a household must integrate their personal backgrounds into some common identity. Kosher practice provides an almost unique mechanism for doing that. While a Jew can attend most religious rituals as an individual, kosher practice requires a family commitment. Not only do families eat together, but the strictures of kosher food preparation require the vigilance of all family members. An individual cannot maintain strict kosher practice unless others take care to keep milk and meat separate, to maintain a rigid division of milk and meat utensils in the kitchen, and to keep proscribed foods out of the house. A careless action by any member—using the wrong scrub brush, placing the wrong dishes on top of one another, mixing the wrong ingredient into a recipe—can defile food for the entire family. Accordingly, working out dietary rules creates a context in which families can and must confront the differences in their understandings of self. It brings individual identity face to face with family identity, and it demands an accord between them.[17]

These differences appear most obviously in mixed marriages, in which the non-Jewish partner generally has little or no prior familiarity with Jewish ritual practice. Yet dietary differences may be just as strong in Jewish marriages, since kosher observance varies so widely. Indeed, since both partners understand the significance of subtle differences in practice, marriages between Jews may offer an even greater opportunity for conflict over food. At time, these conflicts provide a medium for expressing disagreements about Jewish practice more generally. One of my elderly informants found herself frustrated with her husband, who she felt had become too rigidly religious over the years. Unable to get him to discuss the situation, she finally got his attention through the unmistakable gesture of dumping a pot of beef stew onto the milk dishes. At the same time, though, dietary practice offers a vivid way of expressing family solidarity. When speaking to members of mixed marriages, I found most of the non-Jewish partners very interested in kosher cuisine as a way of showing a concern for their mates' culture. They had little interest in religious activities, Jewish or Christian, and in any case they found the synagogue services impossible to follow. By keeping pork out of the house, though, or by learning some of the arcane rules about kosher meats, they could take a visible interest in Jewish tradition. In several cases, non-Jewish wives paid considerably more attention to kosher rules than their Jewish husbands. Kosher practice allows food to become a language through which the commonalities and differences implicit in family life can become manifest.

In many families, kashruth also provides a language for negotiating the difficult issues that surround the movement of children into adulthood. One of my informants, for example, a fifty-year-old architect named Mogens Landau, had given little thought to kosher observance for much of his life; while he and his wife were both Jewish, neither was much interested in religion, and beyond avoiding pork they neither knew nor cared much about dietary law. When his daughter went through bat mitzvah instruction, however, she became very engaged with her Jewish identity, and she decided that she wanted to be more observant. As part of this process, she began lobbying her parents to keep kosher more strictly. Their first reaction was negative—they didn't want to bother with kosher observance, and Mogens bridled at taking orders from his daughter. But within a short time, Mogens said, they changed their minds. After all, what could it hurt? And more importantly, what better opportunity would they have to show their daughter that they valued her opinion? By the time I spoke with Mogens, his daughter had turned sixteen and his family had become strictly kosher. His daughter served as the family expert on kashruth, helping her mother with food preparation and vetting the groceries for acceptability. Mogens was delighted with the change, which he said gave the family something to do together; he was also proud of his daughter's commitment to her principles, happy that she had a mind of her own. I heard similar stories from a number of other families in Copenhagen. Such families had experienced a common crisis in families, the struggle of adolescents to move from the passivity and dependence of childhood to the autonomy and influence of adulthood. Kashruth provided a medium for the expression of that struggle, as well as a mechanism through which families could address it.

Kosher practice, then, allows Jews to express and negotiate the different understandings of Jewishness that pervade social life in a place like Copenhagen. In so small and diverse a community, the creation and maintenance of social groups inevitably involves conflicts between alternative conceptions of self and group. Kashruth translates these alternatives into concrete form, allowing ideas about the nature of Jewishness to manifest themselves as food. It transforms meals into symbolic feasts, statements of group identity that are negotiated and renegotiated three times a day. In different ways, other forms of private ritual observance perform a similar task. Through their mundane intimacy, they infuse daily life with symbolic meaning, and thereby turn the workaday routine of family and community life into an ongoing conversation about the nature of the self.

Conclusion: Religious Practice and Modern Identity in Jewish Copenhagen

The Danish state classifies the Jewish Community of Copenhagen as a *trossamfund*, a "community of faith," a category initially developed to accommodate nonconformist churches in the nineteenth century (Lausten 1987). The word implies a unity of religious feeling, a group of people united around a shared conception of the divine. In its simple sense, this designation is wrong. As we have seen in this chapter, Jews in Denmark do not share a faith, if we understand faith as a particular set of beliefs, nor have they for at least two centuries. The Jewish world includes outspoken atheists as well as devout fundamentalists, observant traditionalists as well as avid reformers, who fight over matters of faith and religious observance much more often than they agree. If anything, variations in observance divide Jews more than shared beliefs unite them. On another level, however, faith does lie at the heart of Jewish identity in Denmark, in that religion forms a common discourse around which the community's factions and disagreements revolve. Even Jews who reject the very notion of the supernatural acknowledge the centrality of religion to Jewish history. Likewise, even non-practicing Jews tend to see Jewish rituals as key moments in the life of their community. Judaism provides a wealth of symbols and practices through which individuals can express their own identities and distinguish themselves from other Danes. Danish Jews do not share their faith the way Lutherans share theirs, taking a common position on key issues of theology; they share their faith the way a family shares its past, some embracing it, some rejecting it, each in a different way, yet none capable of understanding him- or herself except in its light.

This approach to religion has a powerful appeal in a place like contemporary Copenhagen. As Daniele Hervieu-Leger has pointed out, such modern settings have contradictory implications for religious activity (2000). On the one hand, modernity tends to undermine the sorts of communities within which religion has traditionally thrived; the local networks and parochial identities that form the foundation of many churches are shattered by modern patterns of globalization, rationalization, and bureaucratization. On the other hand, these same modern processes create existential dilemmas for individuals that religion is uniquely suited to address. As modernity makes individual identity more changeable, amorphous, and difficult to come by, religion's focus on questions of absolute identity and ultimate meaning becomes increasingly valuable. Religion's ideas become more and more essential, in other words, even as the institutions that embody them become less and less tenable. The difficulty for a contemporary religious group is to supply the symbolic resources its members need so badly, despite the fact that they no longer share local ties or a common worldview.

The ritual world of the Jewish Community of Copenhagen represents one solution to this difficulty. The Jewish Community does not attempt to impose a particular creed or theological viewpoint on its members; it tries, rather, to provide a common ritual structure, a set of symbols and actions which bring together Jews of many ideological persuasions. The MT does not provide a blueprint so much as a toolbox, a set of pre-existing symbols out of which individuals can build their worlds. As they find different ways of engaging with these symbols, Jews make different statements about who they are and how they view the community. What unites them is not the conclusions they come to, but the fund of symbols on which they draw. Whether they reject observance or embrace it, whether they pray in the synagogue or meet their friends there, whether they keep strictly kosher or eat ham for breakfast, they remain Jews, to themselves and to each other, because they take Jewish religious observance as the reference point by which to understand their own actions.

How long such a solution can last is an open question. When I asked Copenhagen Jews about the religious future of the community, they gave me widely differing answers. Some pointed to the steadily declining membership rolls and predicted a slow descent into oblivion. Others, noting the declining influence of the orthodox wing, predicted a Reform movement that would bring disaffected liberals back into the fold. Others talked of a return to religiosity among the young, which might bring back more traditional ritual forms. None of these predictions is inherently implausible, and more than one might well come true. Whatever the future brings for the Copenhagen Jews, however, it is unlikely to bring them unity. Religious fragmentation has characterized the Jewish Community of Denmark since its inception, as the forces of immigration, economic change, and social disruption have repeatedly reshuffled the religious and cultural subgroups which make it up. The world of late modernity has extended this instability to the individual level, forcing Jews repeatedly to rethink and reformulate their notions of self. As long as the Jews of Denmark remain part of modern Danish society, the diversity of factions and viewpoints that finds such vivid expression in its current ritual structure is unlikely to go away. Despite all the forces that threaten it, therefore, the solution to the modern dilemma represented by the Mosaiske Troessamfund has powerful reasons to endure.

3

The Communal World
Jewish Subgroups in Copenhagen

An old Jewish joke begins with a man marooned on a desert island. After six months, he is spotted by a passing ship, and a group of sailors comes ashore to rescue him. He shows them around the island, pointing out his hut, his fireplace, and the pile of rocks that has served him for a synagogue. "After all," he says, "God is everywhere, and even here I need a place to worship Him." "That's beautiful," responds the captain, "but I notice that there's an identical pile of rocks over there. What's that one for?" "Oh, that," responds the man. "That's the synagogue I don't go to."

Like most durable jokes, this one expresses a truth: that Jews tend to form factions, to divide themselves into opposing social and theological groups. Such divisions occur even in very small Diaspora communities, and in large Jewish centers like Israel and New York they can produce dizzying networks of movements, rivalries, and alliances. They certainly occur in contemporary Copenhagen, where the competition among different subgroups shapes almost every aspect of Jewish community life. To say that one is a Copenhagen Jew is something of an incomplete statement, like a sentence without a predicate; being Jewish implies not only membership in the whole community but also an affiliation with one or more of the factions that make it up. While the conflicts among these factions can produce anger and hard feelings, they also create much of the energy and dynamism that characterizes the Copenhagen Jewish world. This sense of internal division, this ongoing friction and turmoil within the group, is as central to Jewish identity in Denmark as the ritual world of the synagogue.

It is tempting to attribute this factionalism to contemporary social trends, to the patterns of immigration and assimilation that have arisen since World War II. Yet divisions have characterized the Copenhagen community since its inception, and indeed since before that. As noted in chapter 1, Christian IV's 1628 invitation of Jewish merchants to Glückstadt specifically mentions oppositions between Sephardic and Ashkenazic Jews (Balslev 1932: 5). In Copenhagen, the legalization of Jewish religious services in 1684 applied only to the German Jews connected with Meyer Goldschmidt; as late as 1695, Sephardic attempts to hold services were shut down by police (Balslev 1932: 15). This opposition receded toward the end of the eighteenth century, but new divisions quickly took its place: among different synagogue leaders throughout the eighteenth century, between reformers and traditionalists at the beginning of the nineteenth, between the rabbi's supporters and disgruntled separatists as the nineteenth century wore on. When the twentieth century opened, Eastern European immigration divided Jewish Copenhagen into two broad classes, divided along lines of language, occupation, and religiosity (Blüdnikow 1986). Shortly afterward, a religious schism split the community again. While one can find a few periods of relative unity in the four centuries of Jewish life in Denmark, these moments appear as rare exceptions to a general pattern of contentiousness.

Such moments of unity do not, moreover, mark high points in the community's history. The late nineteenth century, for example, constituted perhaps the most tranquil period in the history of Danish Jewry; while a few splinter synagogues did exist, the congregation generally enjoyed a placid unity under the leadership of Abraham Alexander Wolff. Such unity did not produce an expansion of Jewish thought, identity, or community activity. On the contrary, it saw the membership of the community fall precipitously, as growing affluence and acculturation led increasing numbers of Jews to intermarry and leave Judaism. Only the subsequent flood of Eastern European immigrants—with all the attendant confusion and discord—halted the Jews' rapid demographic decline. In this sense, it would be wrong to see disunity among the Danish Jews as something inherently dysfunctional, as a dangerous result of a community crisis, or as a harbinger of impending fission. Factionalism and disagreement are basic elements of Jewish life in Denmark, and the community seems to thrive better with them than without them.

This chapter explores this dimension of the Jewish community, the collection of subgroups whose oppositions, alliances, and antagonisms animate the Danish Jewish world. It considers some of the principles along which these smaller groups divide, then briefly discusses some of their histories and subcultures. Our aim is to explore how internal differentiation shapes Jewish life in Denmark and what it means for the nature of the Copenhagen Jewish community.

VECTORS OF DIFFERENTIATION

The divisions that run through the Jewish community in Copenhagen do not follow a single neat pattern. They consist not in a set of discrete and logically equivalent groups but in a hodgepodge of affiliations that sometimes contradict one another and often overlap. They vary considerably in their origins; while some derive from logical tensions in Jewish religiosity or the community structure, others trace their origins to contingent historical events far removed from the local scene. Subgroups also differ in their level of formality and social salience. Machsike Hadas, for example, has a constitution, a membership list, and significant amounts of property, while the group known as "the Polish Jews" consists mainly of an informal social network. In addition, the field of subgroups in the congregation changes over time, as generational shifts and historical developments create new groups and transform existing ones. Given such volatility, any delineation of Jewish subgroups in Copenhagen must be provisional and somewhat subjective; when I asked my informants to list the major constituencies in the congregation, I never received the same answer twice.

One way to make some sense of this disorder is to think in terms of vectors of differentiation, key lines along which Jews in Copenhagen tend to divide. While members of the community might picture the array of subgroups in a variety of ways, they would generally agree on some of the basic oppositions around which the distinctions revolve. Religious orthodoxy, for example, as noted in the last chapter, provides a generally recognized point of contrast among Jewish Danes. In some cases, groups have structured themselves explicitly around this distinction; Machsike Hadas, for example, dedicates itself to the maintenance of Orthodox religious forms in Denmark. In other cases, this opposition forms an unspoken basis for apparently unrelated social organizations. Thus the Jewish Sewing Club (*Jødiske Syklub*) long served as a social club for orthodox women, while more liberal women joined the Women's International Zionist Organization (WIZO). Likewise, one of the community's three day care centers draws children exclusively from orthodox families, while another one draws mainly from liberals. Understanding the way social groups form in the Jewish community, therefore, requires an understanding of religious orientation as a vector of differentiation.

Beyond religious orientation, four other points of contrast bear mention in explaining Jewish subgroups in Copenhagen:

Place and Time of Immigration

Jews in Denmark, like those throughout the Diaspora, define themselves and their families with reference to a history of immigration. The times when their ancestors came to the country, the places from which they moved, and the reasons for which they left form a basic part of Jewish family histories. In my interviews, all of my informants incorporated such details into their descriptions of their families. In some cases, immigration history plays a preeminent role in structuring individual social worlds; recent Polish immigrants, for example, form a distinct and visible subgroup. More often, this background has more subtle effects in defining prestige and social access within the community. People from the oldest Jewish families have an air of aristocracy, and they sometimes told me of how their family ties placed them in the center of Jewish society. Those with more recent lineages tend to see themselves as more peripheral, and they often referred obliquely to the social privileges reserved for the oldest lines. In more or less direct ways, then, the times and places of immigration structure social groupings among the Danish Jews.

When they talk about times of immigration, Jews in Denmark tend to speak of several distinctive periods. Some can trace their ancestry a long way back, to immigrants from Holland or Germany in the seventeenth and eighteenth century. Their family names—Henriques, Goldschmidt, Rée, Melchior, and others—are generally recognized among Danish Jews and carry considerable prestige. Descent from nineteenth-century immigrants has somewhat less cachet, although it still implies a lengthy Danish history and German Jewish origins. More common, and less distinguished, is descent from one of the major waves of immigration that occurred in the twentieth century (see chapter 1). The latest trend in immigration has been from the Middle East, from which a steady trickle of immigrants flowed to Denmark in the 1980s and 1990s. Most Danish Jews can locate themselves or their ancestors in one or more of these immigration periods.

Using these points of reference, people I interviewed tended to distinguish three main groups within the congregation. The term "Viking Jews" (vikingjøder) refers generally to those who immigrated before World War II and whose integration into Danish culture is essentially complete. The term dates back to the beginning of the century, when it was used to distinguish the old German-Jewish families from the new Eastern European immigrants; nowadays, it encompasses the descendents of those immigrants, distinguishing them from more recent arrivals. Viking Jews speak Danish as a first language, have Danish educations, and move easily in Danish society. They constitute a majority of the Jewish community, and include most of its influential members. The term "Polish Jews" (*polske jøder,* or sometimes the derogatory *polakker*) refers to the refugees who came to Denmark from Poland between 1969 and 1971. The Polish Jews have maintained strong group ties since their arrival; in addition to informal friendship networks, they operate active social and intellectual organizations. They tend to speak Yiddish or Polish as a first language, and most are recognizable through their speech and culture as foreigners. Even more recognizable are the Israelis (*Israelske jøder*), recent immigrants who make up about five hundred of the community's members. The Israelis did not arrive in a group, as the Poles did, and their reasons for coming to Denmark are various and largely personal; as a result, they do not constitute an organized community so much as a visible type. While many of them know each other, they do not have the sorts of clubs or political identity within the congregation that the Polish Jews have developed.

Occupation

No dominant occupations characterize the Jews of Copenhagen today; while it would be fair to describe Jews as predominantly middle class, the American stereotypes of law and medicine as characteristic "Jewish occupations" have no parallel in Denmark. They did in the nineteenth century, when commerce and finance drew disproportionately large numbers of Jews, and in the early twentieth century, when Jewish refugees entered the garment industry in droves. Nowadays, however, the occupational structure of the Jewish community differs little from that of Denmark as a whole, and one cannot easily divide the Jews into a small number of occupational categories. Nor can one divide them into classes based on wealth. While some Jews are quite wealthy and some relatively poor, the sorts of upper-class and lower-class subcultures that characterized the community early in this century have long since disappeared. Economic activities and material success do not divide Danish Jews in the broad ways that religiosity and immigration history do.

That said, people in Denmark tend to identify strongly with their occupations, and worklife produces contacts and experiences that shape individual social life. Accordingly, individual occupations often create distinct networks and subgroups. One Jewish lawyer, for example, told me that he knew most of the other Jewish lawyers in Copenhagen; while they had no formal organization, he tried to steer business to other Jews when he could, and he expected others to do the same. A Jewish actress likewise told me that she kept in touch with other Jewish entertainers, all of whom could commiserate about the difficulties and challenges facing such high-profile minorities in Denmark. While informal, these groups often influence the lives of individual Jews much more directly than does the Mosaiske Troessamfund; the lawyer went to the synagogue only once a year, for example, but his Jewish occupational network affected him on a daily basis.

In a few cases, Jews have turned occupational affiliations into formal organizations. Early in this century, for example, a group of skilled tradesmen formed the Jewish Craftsmen's Society (*Jødiske Håndværkerforening*) as a voluntary association within the Mosaiske Troessamfund. The society sponsored social gatherings for craftsmen and their families, as well as informational programs to promote their welfare. The society still exists, although the steady decline of the skilled trades in the community has severely depleted its ranks. In the early 1980s, several Jewish professors organized a similar society for academics, known as "The Academy" (*Akademiet*). The group's fifteen or so members meet every month for a discussion on a topic relevant to Jewish scholarship. Beyond the content of the discussions, the meetings serve to gather Jewish academics together as a social group, giving them a corporate identity otherwise absent in a community that puts little emphasis on intellectual pursuits.[1] Such formal organizations give occupational groups a visible presence in the MT, and they can at times take an active role in the cultural and political life of the community.

Issue-Specific Divisions

In addition to sociological divisions like those mentioned above, groups within the Jewish community also form around specific issues of policy or personality. In some cases, existing oppositions shape the composition of these groups; orthodox and liberal Jews, for example, may line up on opposite sides of a dispute about conversion practices. In many cases, however, issues divide existing subgroups, creating new cleavages and generating new alliances. These affiliations tend to have shorter lifespans and to change more rapidly than others, but they have considerable force while they last, and they have powerful effects on the political life of the MT. Most of these groups never attain any formal status. While members of the MT generally know who belongs to the different camps, the groups seldom surface as voluntary associations or official subdivisions of the community. Congregational politics, for example, often involves sets of allies squaring off against one another, as rabbis or MT officials try recruit networks of supporters to defend or improve on their positions. A particularly bitter conflict broke out in 1982 between the chief rabbi, Bent Melchior, and Finn Rudaizky, the president of the MT (see chapter 4). Both leaders worked feverishly to build support for their positions, and sides became starkly drawn; when I interviewed community members fifteen years later, most would still tell me where they and others had stood on the dispute, and many still nursed a palpable resentment against supporters of the other side. Even in the less contentious period of the late 1990s, my informants frequently described individuals as "a member of Rabbi Lexner's inner circle" or as "a close ally of [President] Jacques Blum." Such affiliations play a large part in the ongoing life of the MT in ways that sometimes cut across religious or cultural subgroups.

More formal groups tend to develop around issues involving the world outside the MT. In 1978, for example, opposition to Menachem Begin's policies in Israel led a number of young Jews in Copenhagen to form a group called New Outlook. In arguing for acknowledgement of Palestinian grievances, New Outlook departed from the MT's generally uncritical support of Israeli policy, going so far as to join Palestinian demonstrators in picketing the Israeli embassy. The group still exists today, although it has receded from its activism and devotes much of its energy to producing a Jewish cultural magazine. Other issue-oriented formal groups in the community include an association to help Russian Jews and a committee to support the Institute for Blind Children in Jerusalem.

Inside/Outside

This last vector operates less as an actual principle of grouping among Danish Jews than as a trope through which group divisions are expressed. The contrast between inside and outside has a very definite and evocative meaning in Danish culture: inside has connotations of warmth, safety, and belonging, while outside implies coldness, danger, and exclusion. This opposition expresses itself in innumerable ways in everyday Danish life, from the tendency to sit in circles at social gatherings to the focus on courtyards and interior spaces in Danish architecture. In social interaction, Danes tend to value membership in the group, being an insider, and to dread being left outside. The opposition has a palpable reality to it, and such experiences as the ineffably Danish feeling of *hygge* are essentially connected to being inside.[2]

This notion of inside and outside resonated with the Jews I interviewed as well. Indeed, I suggested to some, and they strongly agreed, that it was one of the ways in which the Copenhagen Jews were most distinctively Danish. The distinction came up constantly in interviews, as people spoke of themselves as part of the MT's valued inner circle or its neglected periphery. For those who saw themselves as outsiders, this distinction was sometimes a source of considerable distress; they talked poignantly of feeling shut out and isolated, held at a distance by the congregation's insiders. Immigrants and converts felt this exclusion particularly acutely. No definitive criteria establish one as an insider, but a long family history in Denmark certainly helps, as do social connections with the rabbi, attendance at the Jewish school, and active participation in key Jewish organizations. The majority of Danish Jews do not have such traits, and indeed, only a few of my interview subjects claimed to be part of the community's inner circle. Most of them felt shut out in one way or another. Those who did not tended to have a strikingly different view of the MT than the others. They described the Jewish Community as a warm and supportive home, as a group of people who accepted them in a way that other Danes, for all their good intentions, never would. Outsiders, by contrast, often had a strain of bitterness in their descriptions of the MT, which they sometimes characterized as an exclusive and rather snobby club.

The distinction between insider and outsider is not a clear-cut one, hinging more on self-perception than on any objective criterion. Even so, it informs the other, more easily visible differences that the Copenhagen Jews draw among themselves. Occupational networks draw their appeal not only from the practical benefits they confer, but from the sense of being inside the group that they imply. The difficulties facing Jewish immigrants, conversely, involve not only language problems and culture shock but the cold shoulder turned to outsiders in the Danish Jewish world. This dichotomy creates a common experience for peripheralized groups, and it sometimes gives them a basis for common action.

FORMALIZED SUBGROUPS IN THE JEWISH COMMUNITY

As noted above, many Jewish subgroups have no formal structure, and some derive from political affiliations that shift and evolve over time. Others, however, have a strong formal existence that can endure for decades and even generations, and they consequently have a particularly strong role in shaping congregational dynamics. To get a sense of the workings of such groups, let us look briefly at three of them—one group based on immigration, one based on religious orientation, and one new group attempting to establish itself. We will look not only at the structures of the organizations themselves but also at how they fit into the lives of a few representative members.

The Polish Jews

The Association of Polish Jews (*Forbundet af Polske Jøder*) meets once a week in a dreary gray rowhouse in Nørre Farimasgade, the only external indication of its existence a label next to a doorbell reading "Klub." I first went there in June 1997, early in my fieldwork among the Danish Jews. In two weeks of interviews, I had been initiated into the outsider's experience in the community. At the synagogue, the Troessamfund offices, and a number of homes, I had met a succession of lukewarm handshakes, suspicious questions, and carefully restrained conversations. The cold reception had contrasted dramatically with my earlier fieldwork among Lutheran farmers in West Jutland, and the gloomy exterior of the building promised another disheartening encounter. When I pressed the button next to "Klub," a blast of static and incomprehensible Polish responded from the intercom. I had been invited to come by a woman I had met at the cemetery; I shouted her name several times into the microphone, until someone finally buzzed me in. I climbed the dim staircase to the second floor, where a short, bald man of around sixty waited at a half-open door. I told him in Danish that I was an American professor, that I had come to see the club, that this woman had invited me, all of which he took in without a word. Then he stood aside and I walked in, dreading the awkward, suspicious encounter I knew awaited me.

I could hardly have been more wrong. What awaited me was something very familiar, a scene straight out of the homes of my Jewish relatives in New York. Inside the large room sat two dozen men and women, most of them middle-aged or older, drinking hot tea out of glasses at small rectangular tables. Their conversations produced a lot of noise, much more than at an ordinary Danish gathering, with an intensity of gesture and animation that one rarely sees in the Nordic world. Everyone looked up as I came in, and the woman who had invited me waved from her table; I threaded my way over to her and sat down in a free chair amidst four elderly women, promptly burning my hand on a glass of tea that was thrust at me. They soon resumed their conversation in animated Yiddish as I sat smiling vaguely and looking around at the room. Suddenly my hostess burst into laughter, almost falling off her chair, and I turned to see the women all staring at me with big smiles on their faces. They had assumed, reasonably enough, that I was her son, and they were in the process of complimenting her on how big and handsome I'd gotten.[3] As she sat incapacitated by laughter, bent double on her chair, they began pinching my cheeks and firing questions at me in Yiddish, until she finally caught her breath and gasped out an explanation of who I was. They all started laughing then, and soon began asking me questions in heavily accented English—where was I from, was I married, how old were my children, had I ever been to Texas? Was I Jewish?[4] What about my wife? An hour later I emerged back onto Nørre Farimsgade, slightly dazed and well fed, holding the addresses of several people who had invited me to their homes and the address of a family in Austin that I was to visit if I ever got down there. In my field notes that night, I wrote, "At last! After two weeks, I have finally met some real Jews!"

This reaction was exaggerated, of course. As I came to know them better, I found much more warmth among the Danish Jews than I had met at first, and I found important differences between the Polish Jews in Denmark and my New York relatives. But the contrast between the Polish Jews and the larger Jewish community in Copenhagen is still real, a palpable difference of affect and worldview recognized by both parties. The immigrants constitute the largest and most culturally distinctive subgroup within Jewish Copenhagen, and they have ever since their arrival in the late 1960s. Most of the Polish Jews came to Denmark in 1969, at the height of a campaign by the Polish government to drive Jews out of the country.[5] Denmark, like the other Scandinavian countries, offered virtually unconditional asylum for the refugees, and between 1968 and 1971 the country accepted about fifteen hundred Polish Jews. Arriving to widespread sympathy at a time of nearly full employment, the immigrants received a warm welcome and generous support, and they quickly prospered in their new home. Indeed, relations between the Polish Jews and the Danish public became a model of the successful integration of a displaced population, one that subsequent resettlement efforts in Scandinavia have sadly failed to match (see Blum 1982; Blum 1986).

Relations between the Polish Jews and the Danish Jews, by contrast, have been marked by strain and occasional hostility since early on. Frictions between the immigrants and the MT developed soon after their arrival, in large part because of cultural misunderstandings on both sides. Many Copenhagen Jews, including the leaders of the MT, had anticipated a replay of the Eastern European immigration of 1901–1914. They had expected an influx of impoverished, uneducated, desperate refugees, steeped in Jewish culture and eager to engage with Jewish institutions. The new Jews, they hoped, would reinvigorate an increasingly assimilated community just as their predecessors had done. Such a shtetl Jewry, however, no longer existed in Poland in 1969. Most of its members had been murdered during the Holocaust, and the rest had long since emigrated to Israel or the United States. The refugees who arrived in Copenhagen were largely urban white-collar workers, many of them educated professionals or government officials. Few had any deep interest in Jewish religiosity, which had been repressed under the Communist government; some had never considered themselves Jews until they found themselves stripped of their jobs and apartments. Their appearance flabbergasted some of the Danish Jews who first came to meet them. Expecting peasant babushkas in tattered rags, relief worker Hanne Kaufmann was shocked to find refugees wearing stylish dresses and fur-trimmed coats. "They didn't look," she wrote in a widely read memoir, "like what I normally imagined refugees should be" (Kaufmann 1970: 17). She was equally surprised to find them almost totally indifferent to Israel and Jewish religiosity.

Over the next few years, only a minority of these immigrants joined the Mosaiske Troessamfund; many of those who did joined mainly for such practical benefits as a Jewish burial or a Copenhagen residence permit. Their attitudes produced resentment in the MT, where the high hopes for a rejuvenated community went unfulfilled, and representatives of the Polish Jews came in for scathing criticism at MT board meetings. For their part, the Polish Jews bridled at the condescension and criticism of the Danish Jews, who they felt treated them like ignorant peasants. They also resented MT policies that insisted on a halakhic definition of Jewishness. Immigrants with non-Jewish mothers were refused membership in the MT unless they converted, a requirement many found insulting; children with non-Jewish mothers were refused entry to the Jewish school. The result was an antipathy that in some respects continues to this day. In my interviews, Jews born in Denmark often described the Polish Jews as an isolated and inwardly focused community, whose members stuck together and ignored the rest of the Jewish world. Polish Jews, on the other hand, spoke often of the coldness and disdain they felt from the Danish Jews, who called them "Polacks" and thought they were stupid.[6] It was ironic, one told me, that the only place in Denmark he had met people who treated him badly because of his background was in the Jewish community.[7]

If the Polish Jews found little community in the MT, they did find it with each other. The immigrants had not constituted a corporate group before their arrival; they had come from many different parts of Poland, and they often had few or no acquaintances among the other refugees. Once they arrived, however, they quickly formed a close-knit and active community. By 1971 the Poles had established two voluntary associations, one for Polish speakers and one for Yiddish speakers. The groups sponsored lectures, social events, and a cultural journal with articles in both languages. Many of them lived together in an apartment complex on Blegdamsvej, which was later turned into a nursing home as its population aged. Those who joined the MT sponsored Polish candidates for MT offices, ultimately making up a sort of Polish party within MT politics. This community has eroded somewhat in the years since, as the immigrants have aged and their children have become Danes. Today, the two clubs have amalgamated into one, the journal has ceased publication, and community events draw a disproportionately elderly crowd. It is still, however, one of the largest and most active subgroups of Copenhagen Jewry.

The place of this community in individual lives varies. For some Polish Jews, it is virtually their entire social world; not integrated into Danish networks, distant from the Copenhagen Jews, they find among the Poles a familiar and welcoming community. Others live more in the Danish world, regarding the Polish community as a place to return only occasionally, to speak the old language and eat familiar foods for a while before returning to their new lives. This is particularly the case for younger members, who were children at the time of their emigration and who now regard themselves primarily as Danes. Two of the Polish Jews I interviewed, though both of the senior generation, exemplify these alternatives.

Lucyna Rothenberger is a short, friendly woman in her late sixties, with curly gray hair and large brown eyes. A widow and a grandmother, she lives in a small house in the Copenhagen suburb of Virum. She spent much of her life in the provincial city of Esbjerg, in Jutland, where she and her husband were settled when they arrived in 1969. When they first arrived in Copenhagen, the refugee agency put them in a room at the mission hotel in Istedgade, near the main train station; while convenient and cheap, the hotel sat in the middle of Copenhagen's most notorious sex district, and when they stepped outside they found themselves surrounded by prostitutes and pornography shops. This was their whole impression of the city, and when they were offered a chance that afternoon to relocate to Esbjerg, they seized it. Lucyna had been an accountant in Warsaw, and she soon found work in a local library. Her husband, a designer, also found a good job, and they spent several happy decades in Jutland. They always kept in touch with the Copenhagen Poles, however, and when they retired they moved back to the capital. Lucyna has become one of the mainstays of the community, participating energetically in the association, giving lectures at Polish Jewish events, and constantly helping older members who need rides or help with government agencies.

The odds of Lucyna enjoying such a placid and happy life started out extremely slim. She was born in a small town in central Poland, the daughter of a clothing importer, and she was eight years old when the Germans invaded. Her family was split up, and she fled with her father on foot in front of the advancing armies. The story of their escape is among the most harrowing I have ever heard; caught between the two armies, with the borders constantly shifting, with Jews denied passage everywhere, her father bluffed their way through thanks to his facility for languages and an enormous amount of luck. At one point, faced with a phalanx of Polish soldiers blocking the only escape from a killing zone, he insinuated Lucyna through the lines and told her to call to him from beside a house on the other side. As she did so, he managed to convince an officer that he lived in the house, and ought to be allowed home to his family. When I expressed amazement at her story, she said that I shouldn't be surprised at all. "Go down to the Polish Klub," she said. "Ask anyone. They all have stories like this. Because if something incredibly unlikely didn't happen to save you, you died, like everyone else." They eventually reached the Soviet Union, where they stayed until after the war. Lucyna, like her father, returned to Poland a convinced Communist, putting her faith in the system that had protected them from the Germans. The purges of 1968 came as an enormous betrayal to her; to have to flee again, after she had so embraced the new Polish state, seemed epically unfair.

Against this background, Lucyna's experience of Denmark has been a fairy tale. She almost wept when she told me of how generous the Danes were to her, how they welcomed her with a warmth and tolerance that has never been betrayed. She has never experienced anti-Semitism in Denmark, nor even demeaning remarks about her foreignness. When she first got her job in the library, for example, she was put to work cataloging books; later on, when her language improved, she discovered she had made a large number of mistakes. She asked her co-workers why they hadn't said anything, and they said that they hadn't wanted to discourage her. They knew that eventually she'd learn the language better, see the mistakes she had made, and correct them herself. Her husband was a gifted designer, and he achieved a position in his firm that would never have been open to a Jew in Poland. When he retired, the company threw him a lavish party, and he told Lucyna that his years there had been the best in his life. "Can you imagine," she told me again and again, "that a people could be so good to strangers? What other country has ever done something like this?"

I met this refrain frequently among the older Polish Jews, especially those who had experienced the Holocaust directly. Younger people tend to have more mixed feelings, especially those who grew up in Denmark. They point to subtle forms of exclusion that are both more visible and more significant to them than to their parents; they are also, I suspect, more influenced by the pervasive discourse of self-criticism that runs through Danish culture. The older immigrants, though, who contrast Denmark with the Nazis and prewar Poland, seem still astonished by the welcome they have received. Lucyna's mother, with whom she was reunited after the war, came to Denmark soon after she did, and never could get used to the state's generosity. "Why are they giving me a pension?" she kept asking Lucyna. "I never worked for them. I worked for the Polish government for all those years; they owe me a pension. Why is Denmark paying me?" Angry on the Danish state's behalf, she repeatedly badgered Lucyna to get the government to demand compensation from Poland.

Lucyna, like many of the older Polish Jews, has mixed feelings about the Mosaiske Troessamfund. She is a member, she belongs to WIZO, and she attends High Holy Day services; overall, she is probably much more active in MT activities than the average member. At the same time, though, she never feels entirely welcome there. She finds the Danish Jews cold, withdrawn, regarding her as not quite a member of the club. Her occasional contacts with the administration are seldom positive. It is easier to get an appointment with the prime minister, she told me, than with the chief administrator of the MT, and the last time she managed to make one, he stood her up with no explanation. Lucyna does not keep a kosher home, and she is not particularly interested in religion. She is, nonetheless, deeply conscious of her Jewish identity, which she describes as an integral part of her nature. Most of her friends are Polish Jews, and she sees or talks to them every day. If a Jew were to need help, she says, she would offer hers without any hesitation, anywhere, anytime; that, to her, is the core of what it is to be Jewish. As much as she appreciates Denmark and the Danes, they are not her people and do not define her essential self in the way that Jews do.

Lev Nimov has a different view of Jewish identity. Born after the war in Warsaw, he grew up amid the official atheism of Polish Communist culture. His parents were both Jewish, and he knew that he was as well, but they all regarded this ethnicity as irrelevant. They were first and foremost Poles, members of the intellectual class, and they practiced neither Jewish religion nor any distinctively Jewish culture. He studied at the university and became a biologist, working at a research laboratory in Warsaw; he married Elena, another Jewish researcher at the lab, shortly before the anti-Jewish campaign began. Lev described the government's action as simple scapegoating, an attempt to divert attention from its own incompetence. He was accused of Zionism, a charge ridiculous on its face; other Jews faced the same charge, even some who had never previously known that they had a Jewish parent. He told me of one acquaintance who had been told to draw up a list of Jews in his department, which he did in the full knowledge that they would be expelled. When he turned it in, his supervisor praised his thoroughness but reminded him that he'd left off one name—his own, which the supervisor proceeded to pencil in. Lev and Elena moved to Denmark, which they preferred because of its location and its fine universities, and soon found positions at a research institute. Today Lev is a tenured lecturer at the institute, Elena works for a consulting firm, and their two children have recently finished their university degrees.

Visiting Lev's home today, in suburban Hellerup, one sees little evidence of his Jewish origins. He and Elena love Danish design, and the exquisitely modern decor reflects that. Like many educated people in Denmark, they can name the designers of most of their furniture, and they have decorated the walls with posters from favorite museums. No menorahs are in evidence, nor are there any visible mementos of Poland. The Nimovs do not belong to the Association of Polish Jews; they speak well of the group, but they feel at home in Denmark now, and they associate mainly with other people from the institute. They do belong to the MT, even though they never go to the synagogue. After their experience of exile, Lev says, he can see the importance of maintaining a distinctive voice for Jews, and he feels obliged to support the project. Elena belongs to WIZO, through which she attends a social occasion once in a while. At one point, the Nimovs also sent their son to the Jewish school, thinking that he might get a sense of his Jewish heritage there. The boy never felt comfortable there, though, and like his parents and his sister, he does not participate actively in the MT today.

The different conceptions of Jewishness embraced by Lev and Lucyna reflect differences that predated their expulsion from Poland. From early in her life, Lucyna had regarded Jewishness as her defining characteristic; it had shaped her whole social world, including the trauma that transformed her experience of childhood. For Lev, by contrast, Jewishness was an artifact of history, an insignificant feature of his past that became important only when the state persecuted him for it. These views of Jewishness carried over to their lives in Denmark. Once he had settled in a new home, Lev's Jewishness receded again into the background. While he was more conscious of his ethnicity than before, it was the world of his work that defined his identity and social circle. Lucyna re-created, insofar as possible, the Jewish identity she had known before, surrounding herself with the Polish Jewish language and culture so essential to her understanding of self. Neither looked very much to the Danish Jews, who were irrelevant to both of them equally in shaping their understandings of Jewishness.

Things will change in the coming generations. Lucyna's grandchildren, raised in Denmark, have only tenuous sentimental ties to Poland; for them, Jewishness is defined in the context of Danish culture, where the differences between Polish and Danish Jews have little meaning. Their experience of Jewishness differs little from that of Danish Jewish children, and the conflicts and ambiguities of identity typical of Danish Jews will affect them as well. In all likelihood, therefore, the Polish Jews will disappear as a group in another two generations, merged into the larger Jewish community and the larger Danish society. Lev's children are already much like their Danish-born peers, married to non-Jews and distant from the Mosaiske Troessamfund. Born of a historical crisis, nurtured by a common foreignness, the Polish Jews will fade as both that history and that foreignness recede into the past.

Machsike Hadas

If the Polish Jews exemplify the divisions that a new population can generate in the Jewish community, Machsike Hadas exemplifies the ability of old oppositions to endure. Born out of a congregational conflict in 1913,

Machsike Hadas has existed in a state of tension with the larger community throughout its history. The group has established a rich array of institutions that parallel those of the Mosaiske Troessamfund, each of them defined in large part by their contrast with their MT counterparts. In community politics, Machsike Hadas leaders have consistently opposed the MT's central figures, particularly its rabbis. In recent years, as demographic forces have gradually eroded its base, the group has continued largely because of the persistence of some of its most partisan members, who have worked tirelessly to maintain its distinction from the larger congregation. Few of them give the group much hope of long-term survival; its committed membership has dwindled, and only a few related families now maintain its institutions. But for now, and as long as its leaders can keep active, Machsike Hadas constitutes the most active and institutionally engaged subgroup of Danish Jewry.

Machsike Hadas began its existence as a support group for Rabbi Tobias Lewenstein, whose controversial dismissal from the Mosaiske Troessamfund exposed a rift between orthodox and liberal elements of the congregation (for details of this dispute, see chapter 1). Over time, it developed a range of Jewish institutions and began to function as a competitor to the main synagogue. Machsike Hadas could not replicate the entire institutional infrastructure of the MT; it had no cemetery, for example, nor did it have the resources to maintain separate funds for poor relief or care of the elderly. Its several hundred members therefore remained part of the MT, and while they seldom entered the Krystalgade synagogue, they exercised a strong influence on congregational affairs. Given the larger community's fragmentation, for example, a cohesive group like this one could gain significant representation on the MT governing board. Its representatives consistently worked against any liberalization of MT policies, especially with regard to conversion and the recognition of mixed marriages. Orthodox members also set up voluntary associations under the MT's umbrella, sometimes in ways that competed with mainstream associations. With its active, committed, and clearly defined membership, Machsike Hadas exercised influence greatly disproportionate to its small size for almost a century.

The recent decline of the group does not appear to reflect any weakness in the religious commitment of its members. Two other factors, however, have led to a serious loss of membership. On the one hand, the size of Machsike Hadas presented a serious demographic challenge; since its members could marry only other Jews, and since alliances with the larger

Jewish community were problematic, finding enough marriage partners became quite difficult. Members therefore sought husbands and wives outside the country, first in Europe and later in Israel, leading an increasing number of young men and women to emigrate. All of the Machsike Hadas families I interviewed had close relatives living abroad, and all of the young people I spoke with regarded emigrating at marriage as a serious possibility. At the same time, the growing integration and secularization of Danish Jewry after World War II made an orthodox Jewish life in Denmark more and more difficult. Kosher shops closed, Jewish organizations shrank, and the Jewish world as a palpable and encompassing social realm gradually passed away. For those committed to a distinctively Jewish life, other countries increasingly beckoned—especially England and Israel, with their large Orthodox communities and relative proximity to Denmark. From the 1970s onward, a growing proportion of young people in Machsike Hadas began moving away, leaving the group increasingly small, aged, and externally oriented.

The group's corporate existence today derives largely from the work of one man, the elderly industrialist Erik Guttermann. A spry, elfin figure of extraordinary charm,[8] Guttermann has gone to tremendous lengths to keep the orthodox community in Denmark alive. In addition to funding religious institutions in Copenhagen (see chapter 2), he operates a kosher hotel (the Hotel Villa Strand) and a synagogue in the northern resort town of Hornbæk. Given the absence of a viable orthodox population in Denmark, Guttermann has kept services active by attracting Jews from elsewhere—American and British tourists, Russian cheder students, Israeli immigrants, and others. One of my informants recalled a time when his youth group had received a free weekend at the Hotel Villa Strand, and he had looked forward to lazy mornings relaxing on the beach. To his dismay, however, he was awakened early the first morning by his host, who pulled all of the students out of their beds to make a *minyan* for the morning service. (After dressing hurriedly, he jogged over to the synagogue, only to be passed as though standing still by the seventy-year-old Guttermann.) This sort of activism has kept Machsike Hadas going and even given it an intriguing international character. In the long term, however, it seems unlikely to endure. Machsike Hadas no longer exists as a self-sufficient community, and the efforts of Guttermann and a few other key members cannot maintain it forever.

This prospect is a gloomy but inevitable one for Karl Levin, a retired former diplomat I spoke with in 1998. Karl is one of the old Viking Jews, a member of a family that came to Denmark in the eighteenth century, and he is deeply proud of its history. His ancestors have been counselors to the king, famous professors, and important businessmen; they have also been dissenters in the Jewish community, maintaining a family synagogue for much of this and the last century. Karl himself is deeply religious, and he has practiced an orthodox lifestyle throughout his life. When traveling on work, as he often did, he never worked on a Sabbath, nor did he ever eat non-kosher food at an official function. He attends prayer services daily, usually at Guttermann's small basement synagogue. He describes prayers and the Sabbath as the most cherished times of his life, moments in which to feel close to God and suffused with His spirit. His family shares this approach to religion. Karl's wife, an American, works for a Swedish automaker; they have a three daughters, two of whom are married with young children and one who is studying at Copenhagen University.

Karl has always considered himself a part of the Machsike Hadas community. He is a member of its synagogue, and he always celebrates holidays there. The orthodox world does not comprise his entire social circle; he has many non-Jewish friends and colleagues with whom he keeps actively in touch. He stressed to me that his employers always respected his religious observance, making any necessary allowances for it, and he feels fully accepted in the non-Jewish world. Within the Jewish community, on the other hand, Karl does tend to remain in orthodox circles. He has never felt well disposed toward the MT, which he sees as dominated by bureaucrats and secular Jews. The MT administration, moreover, he finds rigid and arrogant. Many years ago, as a young family man, he found himself unemployed and in hard financial straits, so he asked for a temporary release from paying his MT dues. The office responded with an aggressive letter, threatening to revoke his membership and spelling out some grave consequences of failing to pay. Karl promptly quit, and he says that he has never regretted the decision. Nowadays he has next to nothing to do with the large synagogue and its adherents. At Machsike Hadas, he finds a community of friends and relatives who think about religion the way he does, with whom he can experience life as the spiritual journey he knows it to be.

But the community is dying, and Karl's own family exemplifies the reasons why. His wife comes from Brooklyn, where the couple spends a month or so every year. One daughter moved to Israel after graduating from the university, and she does not intend to return; in Israel, Karl says, she can live the kind of fully Jewish life that's just not possible in Copenhagen any more. His other daughters travel to Israel and the United States frequently, visiting family. Now that he is retired, Karl does the same, and if another daughter moves to Israel, so will he and his wife. The thought saddens him; Karl is deeply attached to Denmark, and the family roots and heritage that mean so much to him will have no significance in the Middle East. At the same time, however, Karl is proud of his children's religious commitment, proud that a Jewish life means so much to them, and the thought of living in the Holy Land has considerable appeal. He doubts, moreover, that Machsike Hadas can survive Erik Guttermann for long. It is not only Guttermann's money but also his energy that has kept the community's institutional structure intact—and at his age, he cannot keep going indefinitely. Even now the group has changed, become a collection of old people and foreigners instead of the thriving community of Karl's youth. When Guttermann goes, Karl fears, Machsike Hadas will slowly collapse. And without its community of religious Jews to help him experience God's presence, Denmark will no longer be a place Karl can live.

Chabad House

One of the most intriguing groups to develop in Copenhagen over the past few years is Chabad House, a branch of the International Lubavitch Association that began operation in Denmark in 1996. The Lubavitch Association represents a strain of Judaism that has never before had representation in Denmark: Hasidism, which combines stringent observance of halakhah with exuberant and often ecstatic spiritual engagement. Hasids have become one of the world's most visible Jewish movements in recent years, in part because they favor styles of dress associated with the Jewish communities of prewar Europe. Their men, with black suits, flowing beards, and broad-brimmed black hats, have become icons of Jewry in much of the western world. Their visibility also derives from their astonishing growth, driven by very high birth rates. For many Western Jews, Hasidism has excited both excitement and unease—excitement about the success of a group that neither conceals nor apologizes for its Jewishness, unease about the social and religious conservatism that it promotes. Most Danish Jews I interviewed shared this ambivalence. Some saw in Chabad a promise of a revitalized and newly self-confident Danish Jewry, while others saw the specters of fundamentalist intolerance and Jewish isolation. Either way, almost all of them had an opinion, and most expected the group to play an increasingly important role in community life.

I first encountered the group in the summer of 1996, during my first fieldwork in Copenhagen. In the courtyard of the synagogue after the Saturday service, I noticed a young man who clearly didn't fit in with the rest of the congregants. With his long-tailed black coat, his huge black fedora, and his bushy red beard, he moved among the crowd like a priest at a beach party. I introduced myself, and he asked me back to his apartment for lunch. We walked there, a distance of a mile or so, together with an American tourist and a non-Jewish Dane who was thinking of converting. Over a cold supper, served by his wife, he told us about their mission in Denmark. He was a rabbi, recently finished with his studies in New York, and he and his wife had moved to Copenhagen shortly after their wedding to start a Chabad chapter. Their purpose was to awaken Danish Jews to their Jewish identity. The Danish Jews, he said, were hardly aware of being Jews at all; they didn't follow the commandments, they didn't go to services, they didn't live Jewish lives. He and his wife were to teach them about all that, to put them in touch with their Jewish nature and encourage them to practice. I asked them how long they intended to stay, and he said the rest of their lives, unless the Messiah came first. They were committed to building a community here, and they would do it. Looking at them, I couldn't help feeling a measure of pity. Such young people, just married, expecting their first baby, starting their lives together so far from the communities they had grown up in. They didn't speak the language, they didn't know much of anyone, and the only people they had attracted to their Sabbath meal were two Americans and a Christian from Odense. I wouldn't have wanted to be in his shoes, and I wondered how long he would last.

Four years later, I joined Yitzi and Rochel Loewenthal for another Sabbath dinner. By this time they had moved out of their apartment into a townhouse in Frederiksberg, rented to them at favorable rates by Erik Guttermann. They also had two children, and a third due very soon. The biggest change, though, was the crowd. About twenty people had come for the meal; they had formed a long ragged column on the walk from the synagogue, little clumps of Jews chatting as they walked, following Yitzi's black-clad figure through the city like a band of very well-dressed pilgrims. Some of them were American tourists, but a majority lived in Denmark, and they came to Chabad House fairly regularly. Inside the apartment, a long table stretched across the entire downstairs, set with a cold lunch. We all socialized for a little while, then sat down to a meal that lasted for over an hour. Yitzi sat at the head of the table and presided, leading prayers and directing much of the conversation. He no longer looked callow or disoriented; he was at ease, a host and a teacher, and the people in the room clearly regarded him as an authority figure. They referred to him as "the rabbi," and they listened respectfully when he spoke. Over the next week I visited some of the Torah and Kabbalah study groups Yitzi ran, as well as a Friday evening dinner that lasted until after midnight. All were well attended, with ten to twenty enthusiastic guests. In the years since I had met them, Yitzi and Rochel had clearly created a new social center within the Jewish community, an active and in some cases a highly committed circle of adherents.

This group differed in some important ways from other Jewish social groups in Copenhagen. In the first place, most of its members came from marginal positions within the larger community. Many of the regular attendees were immigrants, either from the United States or more often from Israel. Others included recent converts and recent divorcés, as well as a few who were in the process of conversion instruction. The unhesitating hospitality of Chabad House, its promise to welcome fully any Jew who entered its doors, clearly offered a haven for Jews who had difficulty entering the very closed world of mainstream Jewish social life. Most of them were men; except for the Saturday afternoon luncheons, there were very few women at the occasions I visited, and none who had arrived without a male escort. The gatherings also felt qualitatively different from Danish ones I had attended. They felt, indeed, distinctly American. The apartment had an American feel, jumbled and somewhat thrown together, rather than the cool manicured order typical of Danish interiors. The events took place in English, the only language common to the diverse crowd. And because of the number of non-Danes present, the conversation generally touched little on Danish matters. Americans talked about America and their travels; Danes told Americans where to buy kosher food and how to find tourist sites; Israelis talked about Israel. Such patterns make it hard to predict Chabad's future in Denmark. While the group has grown very quickly, it has drawn mostly on the fringes of Danish Jewry, and it has incorporated relatively little of Danish culture into its workings. Whether it can engage mainstream Jewish Denmark or whether it will remain a phenomenon of the Jewish fringe are questions still undecided.

The regular participants I met at Chabad ranged widely in their backgrounds and interests. Many of them were immigrants, like Aveeram Safran, a heavyset musician in his thirties. Born on a religious kibbutz in Israel, Aveeram grew up in a stringently Orthodox home. At the age of sixteen, however, a tempestuous conflict with his rabbi made him question his faith, and by age twenty he had rejected it altogether. He became an studio musician and lived a secular life in Jerusalem, even marrying a Danish Christian he met at the university. After living in Israel for several years, the couple moved to Denmark. A few years later they divorced, but Aveeram stayed on in the country. Denmark, he told me, was a very good place to be an artist, and he had a lot of friends in the artistic community. He has never returned to orthodoxy; he regards religion as an ethical system, its greatest wisdom the importance of valuing one's fellow human beings, and he doesn't need the synagogue to help him remember that. Even so, Aveeram enjoys coming to Chabad once in a while. He knows all the prayers from his youth, and singing with a group in Hebrew is a rare treat. He also enjoys meeting the other Israelis there. I asked him if he thought that the group had made him more observant, and he said no; on the other hand, he said, maybe it was affecting him and he didn't know it. Who knew? Maybe in another ten years he'd be a Hasid, though he doubted it.

Torben Bamberger, by contrast, was born in Denmark, and has been deeply involved in the Jewish community for most of his life. Now in his late forties, Torben has run a succession of businesses with a Jewish orientation. He currently operates a company that imports food from Israel and Europe. He serves on the board of the Krystalgade synagogue, and he attends services faithfully. His home life, however, has been troubled in recent years. He has been divorced, and his latest business venture has been in difficulties; worst of all, his mother, with whom he has remained close all of his life, has deteriorated physically and can no longer communicate with him easily. During this difficult period, Torben has found great solace in the community at Chabad House. Chabad always offers an open door, a hot meal, and a group of people who are happy to see him. It is a place, moreover, that values the religious faith so important to his understanding of the world. His religion has in some ways contributed to his loneliness; his refusal to marry outside the faith in a country with very few Jewish women, for example, makes remarriage much more difficult. At Chabad, though, his religiosity makes him a leader. His knowledge of the law and the scriptures, together with his business experience, made him a perfect fit for the Chabad House board, where he serves enthusiastically. For all of his devotion to the Copenhagen Jewish community, Torben has lately felt on the margins of its social world; at Chabad, by virtue of his faith, he stands at the center.

Overlapping Memberships, Intermediaries, and Outsiders

Hostility, antagonism, and even open feuds occur regularly among the subgroups of the Jewish community. In some cases, disputes seem to divide the community into opposed camps, as partisans on each side try to mobilize support for their positions. Such oppositions, however, are never total. Links of family, friendships, occupational networks, and more invariably join the two sides; given the small size of the Jewish community and the many vectors through which groups are formed, some people always stand midway between any contending factions. For all the hard feelings between the Polish Jews and the MT, many individuals have connections to both sides. In some cases, the children of Polish immigrants have married native Danish Jews; in others, Polish Jews have taken positions at the synagogue or in Jewish voluntary associations. However deep the divide within the Jewish community, cross-cutting social ties always exist.

Since these ties offer a means of communication and possible reconciliation among factions, the people who embody them often play an important role in congregational affairs. In 2000, for example, the Jewish Community elected as its president Jacques Blum, a sociologist long associated with the liberal extreme of the congregation. Married to a non-Jew, author of a number of works critical of the MT, Blum seemed a very unlikely candidate for the position; certainly, he himself would have laughed at the possibility a few years earlier. Yet for all his liberal leanings, Blum's family roots lay in Machsike Hadas, and many in the congregation saw him as a bridge between the orthodox and liberal camps. Such a bridge had a considerable appeal at that moment in congregational politics. Four years earlier, chief rabbi Bent Melchior had retired after three decades of service. Melchior's political liberalism, as well as his long-standing feud with Erik Guttermann, had made the divide between Machsike Hadas and the larger community unbreachable for many years. His replacement by the more conservative Bent Lexner had led to widespread talk of reconciliation. Many Jews I spoke with regarded Blum as a perfect intermediary—while his record of social criticism gave him impeccable credentials with liberals, he had been raised in an orthodox home and could talk to members of Machsike Hadas as a member of the family. As the oppositions among factions wax and wane, such ambiguous individuals can gain considerable power.

People who stand outside the congregational groupings can also take on a disproportionate visibility in Jewish Copenhagen. Such individuals are rare, since almost all Jews have associations with some of the existing factions. Where they exist, however, they can undertake activities which congregational politics normally rules out. One of Chabad House's greatest successes, for example, has been its series of festivals to mark Jewish holidays. The Loewenthals hold receptions in their home at Hanukkah, Passover, Purim, and other holidays, offering a range of games, craft activities, and didactic exercises for Jewish children. The response has been tremendous—hundreds of Jews attend, from all segments of the congregation, creating some of the largest and most diverse Jewish events held outside of the synagogue. Participants described the festivals to me in glowing terms as a wonderful affirmation of the value of Jewish identity. For all their popularity, though, the success of these events clearly hinges on the Loewenthals' outsider status in the congregation. Yitzi has no official status in the MT, nor does he have a formal commitment to any of its subgroups. He attends services at both Krystalgade and Machsike Hadas. Jews regard the Loewenthals, who speak only limited Danish, as Americans living in Denmark, not as Danish Jews. Accordingly, a festival held at their home arouses none of the usual factional tensions. Liberals and orthodox, Poles and Danes, Melchior supporters and Melchior opponents, all can come together in what amounts to a neutral setting. While this neutrality may not last—Chabad may find itself eventually drawn into congregational politics—it provides, for a while, a unifying moment in an otherwise fragmented community.

CONCLUSION: JEWISH FRAGMENTATION AND THE MEANING OF JEWISH COMMUNITY

When I spoke with Jewish leaders in Copenhagen, the factionalism and fragmentation of the Jewish community almost always came up as a subject of regret. MT officials told me of their frustration with factional infighting, which turned any minor projects that they undertook into extended political battles. Religious leaders decried the bitterness of religious oppositions, ridiculous among a people whom God Himself had chosen as a whole. Leaders of various subgroups fretted about the impact of their oppositions on the young, who might be driven to assimilation out of frustration with their elders' parochialism. Many Jews also worried about the impact on the image of Jews among the larger Danish population. With only a few thousand members, didn't Jews need to hold together? Didn't all this antagonism and factionalization threaten to destroy what was already a small and frail community?

The answer to these questions, I think, is no, because I don't think this is the kind of community that depends on unity for its existence. The Danish Jews do not constitute an organized whole, a group unified in a common social structure or set of institutions. What defines them as Jews is not an embracing social system but a particular understanding of self—they are Jews not because they gain tangible benefits from the community, but because the notion of Jewishness is basic to what they see as the primordial essence of their identities. Their attachment to Jewishness embodies what Herbert Gans has called "symbolic ethnicity," an ethnic affiliation based not on membership in a defined social group but on an identification with the perceived meaning of an ethnic identity (Gans 1979). Symbolic ethnicity makes consensus on group values virtually impossible, since every member formulates an understanding of Jewishness that accords with his or her own experience. As Jews go through their lives differently, as they move in different social circles, as they mold their careers and families in different shapes, they inevitably reach different conclusions about what Jewishness is and how it should be manifested as an institution. The differences of viewpoint that permeate the Jewish world in Copenhagen, then, are not anomalies that threaten to fracture a coherent social system, but are rather a basic feature of the identity on which that world is based.

Indeed, what would threaten such an identity is the sort of unity of which congregational leaders so often dream. If Jewishness in Copenhagen involved a single model of Jewish identity, if Jews agreed on a single answer for questions about what it is to be Jewish, the community would be unable to accommodate the diversity of outlooks generated by their members' diverse experiences. If the Jewish community spoke with one voice, only a small number of Jews would see in that voice the essence of their own sense of self. The majority would turn elsewhere, to other voices, and the Jewish world would gradually dissipate. In a late modern society, where individuals must construct identities for themselves amid a cacophony of networks and attachments, a cohesive and unified Jewish community would ultimately be a moribund one.

The Copenhagen Jewish community consists in no such a unified structure; it is an arena of engagement, a setting in which a variety of understandings of Jewishness can come together, contend, and interact. Its profusion of subgroups and factions provides a wealth of ethnic options, a variety of sites where individuals can reinforce and express their particular views of what Jewish identity. Just as the Krystalgade synagogue offers a toolkit with which individuals can construct their own worldviews, so the plethora of subgroups that collide in the Jewish community offers a menu of notions of self from which individual Jews may choose. The intensity of their collisions is an index of the vitality of the community. When I worked with Lutheran religious groups in northwestern Denmark, I could always tell a dying church by the indifference of its parishioners. If they did not know the details of congregational disputes, if they didn't care about the balance of the building fund, if they had no favorites among the candidates for an open pastorate, then the church no longer figured in their lives. Seldom did I meet such indifference among the Copenhagen Jews. My interview subjects often became emotional when we talked about Jewish politics; as they castigated the "fake Jews" of the liberal left, or the "archaic zealots" of the orthodox right, or the "self-righteous snobbery" of the Viking Jews, or the "inbred isolationism" of the Poles, they showed flashes of anger that left no doubt of the importance they attached to the Jewish community. The subgroups are not comfortable, they are not integrated, they are not a cohesive force for the promotion of Jewish values. They are, however—unlike countless religious groups unified around a common vision that have foundered in the secularizing sea of late modernity—fiercely and unmistakably alive.

The Social World

The Life and Politics of the Formal Jewish Community

On a damp morning in October 1998, a few days before the Yom Kippur service, I drove out to Utterslev Mose north of Copenhagen to watch a recreational soccer match. Like most Americans, I know relatively little about soccer; my Danish friends tell me it is a game of intense strategy and intricacy, and I believe them, but it still looks to my untutored eyes like a bunch of men running randomly around a field and kicking a ball. That morning I was having particular trouble understanding the rules, which involved stopping play and shouting "Off sides" every time it looked as if someone might score. After a while I noticed another spectator, a sandy-haired man leaning against his bicycle near one of the goals, and I walked over to ask him what "off sides" meant. He explained in some detail, while I nodded and pretended to understand. Afterward, we chatted for a few minutes. It turned out that he was a journalist, a reporter for a weekly magazine on his way in to the office. He had seen the game in progress as he rode by. He had stopped in amazement when he saw the players on one of the teams—a mixture of tall, blond, fair-skinned men, obviously native Danes, and shorter, darker, curly-haired men, obviously Middle Eastern immigrants. Given the tensions that have attended the integration of immigrants into Danish society, this seemed to him a very hopeful sign, evidence that the long-promised acceptance of Muslims into Danish society was finally taking place. He thought he might even get a column out of it.

It was with some reluctance that I disillusioned him. This team was not, I said, built on the destruction of ethnic boundaries but on their celebration. It was a branch of Hakoah, the Jewish sports association in Copenhagen, and all of its members were Jews. The differences in their appearance reflected the different lands from which they had immigrated to Denmark, as well as the degree of intermarriage that had occurred among their ancestors. He shook his head in some regret when I told him, seeing his column evaporating, and he said that maybe the match didn't say as much about Denmark as he had thought. On the other hand, he remarked, it might say something interesting about the Jews. Their ability to value religious unity over physical variation was something other Danes could learn from. When he had seen them, he had focused on their differences; they themselves, on the other hand, focused on their similarities, and in doing so they created a common identity that manifested itself in a functioning soccer team.

This notion was heartwarming, I thought, but wrong. The Jews on the team were no less aware of their differences than the journalist. Indeed, they saw differences among themselves that he never imagined. The group included both orthodox and liberal members, each of whom saw the others' faith as a sad deviation from the obvious truth. Some players came from families which had lived in Denmark for centuries, others were children of immigrants, a few had immigrated themselves from the Middle East. These and other differences were the subject of real oppositions within the Jewish world and in some cases involved outright antagonism. The players had come together here not because their faith unified them, but for a more prosaic reason—because they all liked to play soccer, and Hakoah gave them a place to do it. This implied a rather different lesson from the one that the journalist wanted to draw, a lesson in the power of institutional structures. Participation in recreational soccer gave the players an experience of community as Jews that the strictly religious and ethnic dimensions of Jewish life could not. It was not their Jewishness that held them together as a team; rather, it was the team that held them together as Jews.

In many ways, a similar case could be made for the Jewish community of Copenhagen as a whole. As we have discussed in the last two chapters, the Danish Jews are a singularly fractious community, split into an array of cross-cutting cliques, factions, and subgroups. While some Jews are heavily engaged in Jewish religious or cultural activities, moreover, most are not. They experience Jewishness as one part of a larger, and largely Danish identity and social world. Their experiences and what they make of Jewishness differ with their personal histories and philosophies. What, then, makes it possible to talk of them as a community? Where does the sense come from, so evident in speaking with Danish Jews, that they constitute a distinctive group of people? A large part of the answer lies in their institutional structure, the network of voluntary associations, Jewish institutions, and administrative mechanisms that they have in common. For many Jews, especially the less observant ones, these associations and institutions constitute the most concrete ongoing manifestation of their Jewish identity. While they may enter the synagogue only very occasionally, non-observant Jews very often have a regular interaction with a WIZO discussion group, the Jewish school or nursing homes, or a Hakoah sports team. This institutional structure casts a broad net—it allows participation by a wide spectrum of Jews, offering avenues of engagement that appeal to individuals with many different viewpoints on what Jewishness means. All of the structure's branches, however, have connection with a single center, the administrative entity known as Det Mosaiske Troessamfund. They therefore manage to create a common space within which many different social networks and conceptions of Jewishness can come together.

This chapter will explore this dimension of Jewish life in Copenhagen, the institutional framework of its social world. After a brief description of the structure of the Mosaiske Troessamfund, it will examine the three primary mechanisms through which individual Jews come into contact with the organizational world of the Jewish community: voluntary associations, Jewish institutions, and the politics of the MT leadership. As it does so, it will consider the kind of community that these encounters create, and what it means for the nature of Jewish identity in contemporary Denmark.

THE MOSAISKE TROESSAMFUND[1]

The offices of the Mosaiske Troessamfund lie in Ny Kongensgade, a quiet street of tall rowhouses just west of Christiansborg Palace. The street has none of the bustling activity that surrounds the synagogue on Krystalgade; its massive buildings feel faceless and closed, pressed up against the narrow sidewalks with solid locked doors and blank curtained windows. Little distinguishes the MT[2] building from the others, only a modest brass plaque by the door listing the offices and hours. Through the door is a cavernous hallway, and visitors must negotiate two security doors and three flights of stairs before they encounter another human being. When they finally penetrate the reception area, they meet neither the grandeur of the synagogue nor the brimming liveliness of the Polish Klub; they find an office, much like an office at a medium-sized business or a university department, with two busy secretaries shuttling among tables full of paper and office equipment. The rest of the building is a warren of small offices, storage areas, meeting rooms, and a dusty library, most of it empty of people. The building and its few denizens are neither inviting nor warm. After my first visit there, I reflected that I had had a cozier experience applying for my mortgage.

I was not alone. In my interviews with Danish Jews, I seldom met a person who spoke of the MT affectionately. More often I encountered a moderate resentment, a degree of frustration with an institution that in one way or another fell short of expectations. The orthodox referred to it as a den of liberalism and irreligion; liberals tended to characterize it as too orthodox and inflexible; almost everyone found fault with some aspect of its bureaucracy. A few people spun out conspiracy theories, suggesting that its longtime director had maintained his position by blackmailing the chief rabbi with incriminating photos. For all that hostility, however, the MT stands very much at the center of Jewish life in Copenhagen. Almost every Jewish voluntary association has some ties to it. Many have offices there, most use its facilities, and a few receive significant funding from it. The MT supervises and supports all of the key Jewish institutions and serves as the primary point of contact between the Danish government and the Jewish community. Even the MT's harshest critics acknowledge its centrality to Jewish activity, and its policies and politics were subjects of concern to almost every Jew I spoke with.

For an institution with a very small administrative staff, the MT has an impressive mandate. It maintains the Krystalgade synagogue, employing the chief rabbi and the support staff. It performs legally valid marriages, funerals, and divorces for Jews in the city. It represents Jewish concerns to the Danish government, and it keeps official demographic records for Jews in Copenhagen.[3] In addition to its support for Jewish organizations, the MT manages two Copenhagen cemeteries, two nursing homes, and a school with two hundred students. It is also charged with fostering Jewish identity and community engagement in all of their forms. Its reach extends, in one way or another, to almost anyone with an active interest in Judaism in Denmark. It does so, moreover, in a remarkably inclusive way. The MT does not restrict its support of Jewish activities on the basis of ideological or religious viewpoint; its beneficiaries include groups that sharply oppose one another, and even some which the MT leadership opposes. During the early twentieth century, for example, the MT provided meeting space and financial support for Yiddish cultural groups, even though its president openly loathed their leftist orientation (Welner 1965). More recently, I spoke with the organizer of a Jewish literary group who excoriated the MT leadership for its disinterest in the arts. Despite what he described as the MT's scorn for his group, he admitted that it had subsidized it for several years. Such support is affected by political concerns, of course, and many informants recalled examples of MT discrimination against those whom its leaders disliked. Even so, the MT interprets its mission broadly by almost any standard, and the leaders I interviewed evinced a genuine concern with promoting Jewish life in all of its varieties.

Such a mandate requires considerable resources, most of which come from the MT's membership. The largest resource is labor; while the MT employs a few office administrators, voluntary associations carry out most of the work involved in Jewish activities. In addition, like most independent religious communities in Denmark, the MT requires annual dues (*ligningsbidrag*) from its members. Dues vary according to family income, but they tend to be significantly higher than the church tax (*kirkeskat*) which most Danes pay to the state—a fact that members emphasized to me repeatedly in interviews.[4] In the late 1990s, the annual combined dues of the MT's members averaged around seven million Danish kroner, or just over one million dollars. In addition, Jews with reserved seats in the synagogue pay annual rental fees (*stadepenge*), and institutions like the school and the elder homes receive tuition and rents. State subsidies provide an important source of funds to Jewish institutions, especially the nursing homes. Even with government support, however, the MT's obligations are large for a community of a few thousand people, and remaining solvent presents an ongoing difficulty.

Membership in the MT is open to any Jew who wishes to join. Since no reliable figures exist on the total number of Jews in Copenhagen, it is impossible to say what percentage of them belong to the Troessamfund. None of those I interviewed, however, estimated the figure as more than half. Some Jews I spoke with had left the Troessamfund after political disagreements with its leaders; others had objected to the high dues. Others had never belonged, either because their parents had left the group or because they had immigrated from overseas. The absence of such members poses a real threat to the community's finances, and reversing their losses poses one of the most difficult challenges facing the MT leadership. Another problem surrounds the status of intermarried Jews. Following the halakhic formulation, the MT defines a Jew as a person with a Jewish

mother or a person who has converted to Judaism. The child of an intermarried Jewish man, therefore, must undergo conversion in order to become a member of the MT. Given the high rate of intermarriage, this
situation covers a significant number of people, and it creates difficult pro-
cedural questions. Should the children of intermarried men, for example,
be admitted to the Jewish school? Should they be allowed to join Jewish
youth groups or to attend Jewish social functions? Many orthodox Jews say
no; admitting such children would undermine the halakhic definition of
Jewishness, and it would present intermarriage to the young as an approved
custom. On the other hand, rejecting them often offends their parents, and
it makes it much less likely that the children will later convert. The result,
say many liberal Jews, is a further decline in an already dwindling population. The boundaries of the Jewish community, and the status of those who
fall just outside of them, make up a contentious and ongoing subject of
dispute within the MT.

Most Danish Jews enter the Ny Kongensgade offices fairly seldom; unless they have problems involving their membership fees, they have little
cause to interact directly with the MT bureaucracy. Their engagement
with the institution usually occurs indirectly, through one of the activities
or institutions it sponsors, or through the politics surrounding its leadership. Let us turn now to a more direct look at these points of contact, beginning with association life.

VOLUNTARY ASSOCIATIONS IN THE JEWISH COMMUNITY

When I did my dissertation fieldwork in 1990–1991, I lived in a small market town on an island in northwestern Jutland. The town had about 8,500
residents, who lived clustered around a small port on the Limfjord; it had
two supermarkets, one movie theater, perhaps a dozen restaurants, and one
main church. What it had in profusion, however, were voluntary associations. The local address book listed over 190 voluntary associations, and at
least as many more existed without a listing. The town was by no means
unusual. Danes love voluntary associations, groups organized to promote a
specific social or political purpose, and most belong to at least a few of
them. Associations range from national campaigns like the Cancer Society
(*Kræftens Bekæmpelse*) to collections of friends who meet to play cards or
knit. They tend to have a relatively formal structure, such that even a small
hobbyists' club might have a written constitution, a board of officers, regular fixed meetings and a dues schedule. A surprisingly large number of

them document their existence; one of the delights of historical research in Denmark is the multitude of anniversary volumes and microhistories that detail the lives of voluntary associations. This pattern characterizes the Danish Jews as well. Much of Jewish life in Copenhagen revolves around voluntary associations, of which at least a dozen list their activities in the community newsletter each month. They span a broad range of purposes, from political organizations to sewing clubs, with new ones cropping up every year or so. Most have the organizational formality typical of Danish voluntary associations—they have board meetings and elections, membership dues, constitutions, and often newsletters or websites. Most also have the same duality of purpose as non-Jewish associations, combining a sincere commitment to a particular goal with the creation of a social network. For many Jews, associations stand at the center of their Jewish social life. They offer a chance to meet other Jews, to focus directly on Jewish subjects, and to talk directly about the complexities of life as a Jew in Danish society. While I spoke with a number of Jews who disavowed any interest in religion, and many who had serious misgivings about the MT as an institution, I seldom met one who had no connection with a Jewish voluntary association.

With a few exceptions, the MT administration does not start voluntary associations. They are organized by their members, who approach the MT for sponsorship as they find it necessary. The number and orientation of the associations therefore varies over time, as different issues and community subgroups rise and decline in prominence. In recent years, voluntary associations in the MT have fallen into several broad categories. Among the largest is Israeli support groups, like the Danish Zionist Association (*Dansk Zionistforbund,* or *DZ*), the Danish Collection for Israel (*Det Dansk Israelsindsamling,* or *DDI*) and the Danish branch of the Women's International Zionist Organization (always referred to as WIZO). Some of these groups, like the Friends of the Institute for the Blind (*Blindeinstittutets Venner*), focus on specific Israeli institutions; some focus on Israeli culture and politics, like New Outlook.[5] Another group of associations focuses on Jewish culture and history, including the Danish chapter of the B'nai B'rith. Some of these put out publications—the Danish Jewish Historical Society, for example, puts out an annual journal called *RAMBAM,* while a literary studies group published a journal called *Alef* in the 1990s. Another type of association focuses on specific political causes, like the Action Group for Jews in the Former Soviet Union. Still another supports the interests of particular groups within the community, such as the Polish

Jews' Association and the Jewish Craftsmen's Society. Some organizations conduct recreational activities; these include youth groups, like Jewish Youth (*Jødisk Ungdomsforening*) and the Younger Jews' Association (*Yngre Jøders Organization*), as well as mixed groups like *Hakoah* and the *Jizkor* choir. Finally, a number of groups support MT logistical activities. The Guards' Association (*Vagtforening*) organizes security for MT events, for example, while a voluntary organization edits and distributes the community newsmagazine, *Jewish Orientation* (*Jødisk Orientering*).[6]

The dual role of association life appears perhaps most visibly in WIZO. WIZO's formal purpose is to raise money and support for Israel. The Danish branch directs its fundraising to the support of several specific Israeli day care centers. Its gathers most of its funds through an annual bazaar that sells a variety of items made or donated by the members. The bazaar raises a significant amount of money, and it marks one of the major yearly social occasions in the Copenhagen Jewish community; it is widely advertised, heavily attended, and it involves months of planning and organization. For all its financial significance, however, the bazaar's real importance lies in the connections it establishes among its members. WIZO divides its membership into small groups, each of which devises a fundraising activity for the bazaar each year. The group meetings provide occasions for Jewish women to get together, socialize, and discuss some of the difficulties of living a Jewish life in Denmark. As one woman put it, "We're supposed to be raising money, I know, but mostly we sit around and gripe." The WIZO leadership works carefully to foster this sort of connection, assigning women as much as possible to groups of the same age, education, and interests. A college student I interviewed met regularly with a group of seven other students and recent graduates, while a Polish Jew met with nine other Poles. Immigrants often find themselves together in a group; one woman told me that her group referred to itself as "the Tower of Babel (*Babeltornet*)," since there were five different native languages among the ten of them. As a result, the association draws from almost all subgroups within the congregation. I rarely met a Jewish woman who didn't belong to WIZO, and most of them seemed to value the moment of Jewish community that the meetings provided. Discussion groups, moreover, offer an opportunity to discuss women's issues in a community that many perceive as dominated by men. Under the pretext of a fundraising operation, WIZO effectively creates a network of women's social circles that spans the Jewish community.

WIZO's structure cleverly solves one of the most difficult problems facing MT associations: attracting a broad range of participants. Voluntary associations tend to recruit through personal networks, and much of their appeal lies in the intimate social circles they create. Accordingly, voluntary associations tend to reinforce existing social relationships rather than create new ones. Remarks about cliquishness and exclusivity recurred constantly when my informants discussed association life; over and over, particularly when a publication was involved, members complained that association boards cared more about appointing their friends than about the groups' official goals. And when I visited voluntary association meetings, I often met an almost palpable similarity of outlook among their participants. In most cases, associations do not create new connections among previously unrelated people, but rather they provide a context in which existing relationships can manifest themselves concretely.

This role does not mean that voluntary associations fragment the community further, because the group lines along which associations take shape are not mutually exclusive. In most cases, an individual belongs to a number of social categories, each of them relevant in different contexts and none perfectly replicating any other (see chapter 3). The membership of a woman's WIZO group differs from that of her Jewish literature club, which in turn differs from that of her old youth group. Some groups, moreover, do have an appeal that enables them to extend beyond limited social networks. The lure of soccer brings people from all parts of the community to Hakoah, for example, while youth group activities draw students from a variety of backgrounds. As a result, while the membership of particular associations is often quite limited, the association system as a whole creates a common field of social interaction among Copenhagen's Jews. When I visited an association meeting, and the name of a Jew outside the group came up, I could always count on someone in the group to know the person—in most cases, because both had served together on one or another association board. Association life creates not so much a single group as a web, a set of networks through which most of the city's Jews, in one way or another, can find a connection with each other.

Jewish Institutions

If the entrance to the MT offices seems unwelcoming, the entrance to the Jewish school is positively intimidating. The school lies in the suburb of Hellerup, its back hard up against the main motorway leading north from the city. It began its life as a textile factory, and it still has that aspect today—a meandering one-story structure of dingy yellow brick, the low roof relieved by triangular eruptions of skylights. An asphalt courtyard runs between the buildings, and a large sandy playground stands off to one side. What dominates the view from the street, however, is the fence. Made of chain links, about twelve feet high, the fence completely encloses the school grounds; for those tempted to scale it, a double row of barbed wire stretches along its top. An enormous gate of steel bars, itself perhaps ten feet high, closes off the entrance to the courtyard. A smaller gate stands just to its left, where pedestrians can seek entrance using a call box under the eye of a pole-mounted security camera. When I first saw the gate, it reminded me of a prison or a military base in the United States, and I despaired of ever getting inside. In a remarkably peaceful country, where schools are usually quite open to the public and where tourists can walk beneath the windows of the royal palace, the security surrounding the school is startling.

The purpose of the gates, on one level, is the prevention of terrorism. In the 1980s, a Palestinian group exploded bombs outside two Jewish institutions, and the MT has maintained a tight security apparatus ever since. All of its buildings have impressive security systems, including strong fences and extensive video surveillance. Beyond this purpose, however, the security installations do something arguably more important: they create a boundary, firm and difficult to penetrate, between the Jewish space within them and the non-Jewish space without. In doing so, they make the Jewish community something concrete, something that involves a physical territory with limits and contours. This feature goes a long way toward explaining the central importance of MT institutions in Jewish Copenhagen. In a diffuse and amorphous community, whose nature and boundaries are a subject of ongoing dispute, Jewish institutions provide a place in which Jews can physically come together, where they can experience their community as something palpable. At the same time, they create a visible set of symbols through which Jews can negotiate questions about the nature of group identity. Institutions provide a place, in effect, for Jews to come together— and, once together, to argue about what they're doing there.

The MT operates a number of social institutions, most of them focused on some aspect of social welfare. Some of them provide services for children; in addition to the school, the MT operates three daycare centers (*børnehaver*) and a Jewish teaching resource library. Others focus on the elderly. These include two nursing homes, the Meyers Minde complex next to the synagogue and the N. J. Frænkels Stiftelse in Utterslev, which together provide care for fifty-four residents. The Raoul Wallenberg Apartments on Nygårdsvej provide another thirty-eight residences for elderly Jews, and the MT Day Center (*Dagcenter*) offers activities for seniors in Ny Kongensgade. Finally, the MT administers the two Jewish cemeteries in Copenhagen, and it maintains others elsewhere in Denmark. Much of the funding for these institutions comes from the state, particularly for elder-care services. The nursing homes represent a partnership between the state and the MT, one which some have suggested may serve as a model for other ethnically based social services in an increasingly diverse Denmark.

These institutions draw participants from all parts of the Jewish community. The school, for example, offers the only Jewish educational environment in Denmark; this appeals not only to orthodox parents, who want to emphasize Judaism throughout their children's experience, but also to liberals, many of whom feel unable to convey a sense of Jewishness in their own homes. Members of old Danish families often have traditions of patronizing the school, while immigrants often find its familiar symbols comforting in their new environment. As a result, a visitor to the school meets children from a broad cross-section of the Jewish population, and meetings of the parents' council are some of the most encompassing events in Jewish Denmark. Likewise, the central location of the Meyers Minde nursing home and its proximity to the synagogue give it an appeal to Jews of all stripes. Much like the Krystalgade synagogue, the community's social institutions bring the diverse faces of Danish Judaism into a single common space.

Unlike the synagogue, however, and unlike association life, social institutions involve sharing that space for a long time. Students at the Jewish school stay together from morning until late afternoon, all year long for up to nine years. Residents of the elder homes live together even longer, sharing meals three times daily. They must therefore work out not only decisions about ritual protocol or the minutiae of association policies, but also the myriad details of everyday life. In the Jewish nursing homes, for example, how strict should rules be about Sabbath observance? May individual members watch television or use hair dryers on Saturdays, or should the entire community adhere to Orthodox prohibitions? In the Jewish school, should boys wear skullcaps (*kippot*) in class? Must students who eat pork at home keep kosher in school? Should students be allowed to have non-Jewish romantic attachments, and if so, should such partners be allowed to attend school functions? These questions imply questions about the nature of Jewish identity and the contemporary meaning of Jewish law, questions that different community subgroups answer very differently. In much of its activity, the MT allows these different conclusions to coexist;

social institutions, however, require a common answer, some joint formulation that all groups can quite literally live with. The urgency of daily existence demands that the community's diverse segments find a common approach to Jewish living.

Doing so is not easy. At Meyers Minde, for example, all residents eat in a common dining room, since individual kitchens for the apartments would be dangerous and prohibitively expensive. Sharing their meals means that residents must practice a common style of dietary observance—a person who keeps kosher cannot use the same dishes or kitchen facilities as someone who does not. The home therefore keeps a very strict kosher regime. Meyers Minde uses only foods certified as kosher by the chief rabbi. Its kitchen facilities are divided into milk and meat areas, and all pots, pans, cutlery, and dishes are color coded for milk or meat. The administration defends the practice as the only reasonable arrangement—any lesser degree of stringency would keep strictly kosher Jews from eating at all, whereas nothing prevents non-kosher Jews from eating kosher food. Yet for some liberal Jews, this system comes as a hardship. The standard Danish diet relies heavily on pork, especially at festive meals, and it frequently mixes meat and dairy products. For Jews accustomed to this diet, the Meyers Minde food seems thin and bland, a disappointment in the already attenuated sensory world of a nursing home. Several people I spoke to seemed angry about this. As one put it, "Here I am, eighty years old, and I have to let some fundamentalist from Machsike Hadas tell me what to eat!" Sharing institutional space, then, can produce both integration and conflict. Joint living arrangements force Jews to agree on a set of rules that bind them all equally; at the same time, they can create an ongoing source of irritation, a reminder of the conflicts that rules embody. This irritation cannot be laid aside, moreover—built into the fabric of life, disagreements over Jewish identity spring to fresh life with every unsatisfying meal.

The bitterest disagreements tend to revolve around issues of membership—who does and does not qualify for inclusion in the Jewish community. Nowhere does this question create stronger feelings than where it touches on the cemetery. While the MT oversees maintenance of several cemeteries, only one currently has room for new burials, the Jewish Cemetery (*Mosaiske Kirkegaard*) next to the massive Western Cemetery in Valby.[7] The cemetery offers its services only to MT members; Jews who have left the MT, or who have never joined, may be buried only upon payment of a substantial fee. More importantly, non-Jews may not be buried there at any price. Most conservative Jews I spoke with regarded this policy as self-evident. The cemetery, they said, is a Jewish cemetery, and only Jews belong there; to admit others would violate the sanctity of the space, setting crosses and other Christian symbolism in the middle of a Jewish holy site. Given the limited space available, moreover, it might create serious problems of overcrowding. For many liberal Jews, on the other hand, the policy flies in the face of their daily reality, which is lived in close association with non-Jewish Danes. For many of them, especially intermarried Jews, their bonds with non-Jews are as deep and intimate as those within the Jewish world. The refusal to bury non-Jews in the cemetery, therefore, appears as archaic and cruel, a policy that divides partners in death who have spent whole lifetimes together. In my interviews, a number of liberal Jews spoke very angrily about this practice. They recounted stories of children who could not bury their parents together, and of widows crushed at the realization that they would be buried far away from their dead husbands. The MT has tried on occasion to address this concern, burying intermarried Jews next to the wall and burying their spouses in the Christian cemetery on the other side. Such accommodations are rarely possible, however, and few find them satisfactory in any case. By creating a localized physical community of the dead, the cemetery exposes the conflicts among conceptions of the living Jewish community and the virtual impossibility of reconciling them.

These tensions also characterize the MT's most active social institution, the Jewish school. The school's official name is the Carolineskole, a reminder of its origins as a girls' school during the early nineteenth century; founded to teach Jewish girls to read Danish, it received the patronage of Prince Frederik's sister Caroline. Its mission has changed considerably since its founding. Established to enable Jewish children to participate in a Danish social world, it now tries to foster Jewish identity and culture among otherwise Danicized students. The Carolineskole has about 200 to 250 students, which most observers estimate as about half the school-age population among Copenhagen's Jews. It offers the equivalent of the nine-year basic education in the Danish state school system, as well as a kindergarten/daycare center (*børnehave*) for the very young. After age sixteen, students continue their studies at the academic, business, or technical schools of the state system.

The Carolineskole's mandate—to offer a full education to any Jewish student who desires it—presents it with some serious administrative and economic challenges. It must maintain a full standard curriculum as well as religious instruction, despite a relatively small and unpredictable population base. Class sizes fluctuate significantly from year to year; one of the classes I visited had twenty-six students, while several others had as few as a dozen. It must compete, moreover, with a state educational system widely regarded as one of the world's best. As a result, despite rather high tuition rates, the school constantly struggles to cover its costs, and at times it has flirted with financial collapse. The strain shows in the condition of the grounds. While the kindergarten has a wonderful playground, most of the school feels drab and slightly unkempt. Classroom furniture is old and chipped, lawns look ragged, and the hallways feel dark and cavernous. Compared with the cheery, manicured schools I had known in Jutland, the Carolineskolen felt rather grim in my first visit there, hardly a place to entice one into Judaism.

And yet it does. For all its difficulties, the school has a very engaged and loyal student body, and one extremely conscious of its Jewish identity. I spoke with a number of current students, all of whom described the Carolineskole in glowing terms. They said that they felt comfortable there, that they felt special, that they could explore and experience their Jewishness there in a way impossible elsewhere. They all spoke a great deal of the centrality of Jewishness to their identities, and many expressed plans to emigrate to Israel as adults. To some extent, of course, these statements simply reflected the school ethos; in a setting where teachers constantly reiterate the importance of Jewish distinctiveness and togetherness, student descriptions of their experience are likely to reflect these themes. The curriculum has a strong Zionist emphasis, capped by a class visit to Israel in the final year. At the same time, though, student enthusiasm for the school was clearly genuine, and parents frequently told me how much their children liked the school. They had a particularly strong relationship with Bent Lexner, the MT's chief rabbi, who teaches religion there. Lexner often describes teaching as the most important part of his job, and indeed he maintains his office at the school rather than the synagogue. The work clearly suits him. Watching him with the students, all of whom he knew by name and genuinely seemed to like, I saw in Lexner a liveliness and warmth very different from his rather stiff presence in the synagogue. While appraisals of the school's academic quality differ, its ability to foster a sense of Jewish identity is agreed on by all.[8]

Along with that identity, however, comes a familiar set of tensions. The same divisions that characterize the larger Jewish community recur in the Carolineskole, notably the painful opposition between insiders and outsiders. When I interviewed recent graduates of the school, they spoke a great deal about cliques—the extent to which students formed closed subgroups, and the extent to which they themselves had felt shut out of the inner circle. The school also raises familiar issues about the meaning of Jewish identity and religiosity. For all their admiration of Lexner's enthusiasm, some parents found his view of Jewish identity disturbing: his emphasis on the importance of religious observance and on the primacy of Israel in Jewish life clashed sharply with their own ideas. Many felt that he had pushed the school in a much more conservative direction over the past decade, and I was told that this had caused some parents to remove their children from the school. The school's emphasis on Jewish distinctiveness also bothers many parents, who see Jewishness as something integrated with a Danish national identity. As one put it to me, "I get sick of all the navel-gazing, of looking at the whole world in terms of how it affects Jews and Israel. My daughter is Jewish, yes, but she's also Danish, and they never want to acknowledge that."

As with the cemetery, some of the most intense disputes surround the issue of access. Who should be allowed to attend the school? In particular, should the children of intermarried Jewish fathers be admitted? Current policy allows them to attend until age twelve; at that point they must formally convert to Judaism or leave the school. Some orthodox Jews I interviewed regarded this policy as preposterous. As one put it, "These children are not Jews, so what is the point of their being in a Jewish school?" Allowing them in, he told me, produces a host of problems: it tells students that non-halakhic Jews are part of the Jewish world; it places the MT's seal of approval on mixed marriages; perhaps worst of all, it may result in marriages between these non-halakhic Jews and real Jews, drawing yet more Jews out of the community. Many of my liberal informants found this reasoning profoundly offensive. They argued that the school should admit the child of any MT member—how could the MT, they asked, justify taking a man's membership dues but not allowing his child into its school? In a community that is losing membership, moreover, and in a school desperate for enrollment, what sense does it make to reject people who want to become part of the Jewish world? This disagreement, of course, echoes a broader conflict between the orthodox wing and liberal Jews over the nature of Jewish identity and the contemporary meaning of Jewish law.

How much this and other conflicts affect student experience is hard to say. Students certainly enjoy the school, and a significant number of them do emigrate to Israel when they complete their educations. At the same time, though, I was struck in my interviews by the tendency of Carolineskole graduates to describe their graduation as a moment of liberation, a welcome release into a larger social world. Most of the school's students continue their education elsewhere, almost always in a much larger school with very few Jews. I asked one man, now a middle-aged businessman, what the change was like, and he fairly glowed as he said, "Fantastic!" What, I asked him, was so fantastic about the Frederiksberg gymnasium? His reply was instant: "Non-Jewish girls!" When I laughed, he explained that he didn't mean to imply anything bad about Jewish girls. "But you have to understand," he said, "what it's like spending nine years with the same little group of people. You know them all, you know their parents, you know them better than you really want to. And the thought of getting married to any of them is just horrible!" Other informants made similar remarks, about feeling released from the little world of cliques and squabbles, where one's Jewishness was always on display and one's commitment to Israel always in need of restatement. Once they leave that world, most of them never return; they move into a Danish social world, they marry Danes, and Jewishness recedes to a single facet of a larger and more complicated identity.

This pattern points to a telling aspect of Jewish social institutions in Copenhagen: their marginality. For all their importance to community discourse, these institutions do not bring together adults in the prime of life, nor do they gather families together in groups. They serve Jews at the margins of life, Jews who are unable by reason of youth or infirmity to fend for themselves. No apartment buildings exist to house Jewish families; no schools provide instruction for Jewish adults; there are no Jewish recreation centers, union halls, reading rooms, or neighborhoods. In part, this absence reflects the influence of state subsidies, which are directed primarily at children and the elderly. It also reflects, however, the nature of community among the Danish Jews. Jewish identity in Denmark is largely symbolic, a matter of self-understanding rather than an extension of a self-contained corporate group. Jewish space is an important resource for the community, a place where Jews can come together and where ideas about Jewishness can meet and contend. It is not, however, a place where most Jews want to stay. As they construct their individual lives, Jews want very badly to take part in the larger world of contemporary Denmark. In the last analysis, strictly Jewish space is populated almost entirely by those who have little choice—by the very young, the very old, the very sick, and the dead.

Electoral Politics in the Mosaiske Troessamfund

At some point, disputes over schools, nursing homes, and other Jewish institutions require some resolution. The MT makes these decisions through a democratically elected leadership structure. The system involves two levels: the twenty-member Congress of Delegates (*Delegeretforsamling*, usually referred to as the DF) and the seven-member Board of Representatives (*Repræsentantskab*, or RS). Election is indirect; members of the MT vote for the delegates, and then the delegates vote for the representatives. While policy authority formally rests with the DF, real administrative power lies in the RS, and particularly in its chair (*formand*, sometimes referred to as the MT president). The president makes the final decisions about issues of institutional policy, and he oversees daily administration for the Troessamfund. These range from minor matters of funding for associations to such major concerns as negotiating the contracts of rabbis. The president in effect personifies the MT as a social entity, and the position carries considerable prestige. Several people told me the story of jurist C. B. Henriques, who was once appointed to a position with the Danish Supreme Court and the presidency of the MT on the same day. Of the two accomplishments, his mother bragged about the MT presidency first.

These stakes make the quadrennial DF elections major events in the life of the MT, events accompanied by extensive and energetic political activity. Much of this activity takes the form of creating and promoting delegate slates, or lists (*liste*). Individuals rarely seek election to the DF by themselves; in most cases, a number of candidates will form a list, which runs for election as a unit. MT members vote in favor of particular lists, and seating in the DF reflects the proportion of the total vote each slate receives. A list that received half of the votes in the election, for example, could place ten of its members on the twenty-member DF. The number of lists varies from election to election, but usually runs between three and six.

In many cases, lists reflect particular subgroups or social circles within the congregation. One list, I was often told, consisted of friends and allies of Rabbi Bent Melchior; another consisted of Machsike Hadas members; another represented the Polish Jews. Such lists provide political leverage for congregational subgroups, allowing them to vote together for blocks of delegates. Other lists focus on specific issues, and tend to arise and disappear more rapidly. The Peace List (*Fredslisten*), for example, arose shortly after the Israeli invasion of Lebanon in 1982. Founded largely by leftist academics, the list called for peace both in the Middle East and in what was at the time a bitterly divided MT. In 1968, the burgeoning youth movement found representation in the aggressively iconoclastic Youth List (*Ungdomslisten*). More recently, in the late 1990s, a group of intermarried Jews established the Tolerance List (*Tolerancelisten*), which gained four seats with its call for a greater incorporation of mixed couples into the congregation. The relative power of the different lists rises and falls with the issues and constituencies they represent. When I visited Copenhagen in 2000, six different lists shared seats in the DF. They included two liberal issue-oriented groups, including the Tolerancelisten; a centrist list affiliated with some of the MT's older families; a list associated with Machsike Hadas; a list associated with the Polish Jews; and a list consisting of a single orthodox activist trying to start a new movement.

The association of lists with community subgroups gives MT elections a significance beyond their impact on institutional policy. For members of the Troessamfund, the outcome of the election indicates not only the makeup of the new administration but also the makeup of the community. The ascendancy or decline of the liberals, the orthodox, the insiders, the outsiders, the immigrants, the young, the intermarried, and all of the other varieties of Copenhagen Jewry finds an oblique expression every four years in the elections. For a person familiar with the Jewish world, who knows the cliques and interests represented by the names on the different lists, the results provide an instructive snapshot of the state of the Jewish world. As a result, the political process receives widespread attention in the Jewish community, attracting the interest even of Jews who have little contact with the MT and little concern with its policies.

The Role of the Chief Rabbi in MT Politics

Many of the political issues that divide the MT have limited lifespans. The Youth List, for example, petered out along with the youth movement it represented, while the changing face of the Middle East conflict eventually undermined the Peace List. Some issues, though, derive from the nature of Jewish identity in Denmark or the structure of the MT, and they persist in strikingly consistent form over time. The conflict between liberal and orthodox interpretations of Judaism provides a recurring source of conflict; the same disagreements that sparked the founding of Machsike Hadas in 1913 remain fresh in the year 2001. The concern about mixed marriages that spawned the Tolerance List likewise existed in very similar form in the years just after the war, or even in the nineteenth century. These issues are implicit problems for any religiously based minority group, and they seem likely to persist as long as Jews remain in Denmark. Less obvious but equally enduring are some conflicts that derive from the administrative structure of the Mosaiske Troessamfund. To give a sense of

MT politics in action, I would like to briefly explore one of them: the issue of the authority of the chief rabbi.

The chief rabbi of Denmark is, officially, an employee of the MT. He is hired by the Board of Representatives to carry out and oversee the religious activities of the congregation; he has final authority over all questions of ritual practice and no power over other congregational activities. In practice, however, his influence extends well beyond the ritual realm. His presence in the synagogue makes him the most visible face of the Jewish Community. He appears frequently in the press and at public occasions, where his views are widely understood as representative of Jewish opinion. Within the community, his roles in ritual and education bring him close to almost every Jewish family; every Jew who has buried a parent in the Jewish cemetery, or held a circumcision, or studied at the Carolineskole has had an extended personal interaction with the chief rabbi. The interpenetration of the ritual and social spheres among Jews, moreover, gives the rabbi authority over areas that might seem outside the religious realm. The boundaries of the group—for example, who may and may not be considered a member of the MT—lie substantially under the control of the rabbi, since he may decline to perform such ceremonies as conversion, marriage, circumcision, and burial. Similarly, since food purchases by MT institutions must meet with his approval, he can sometimes direct patronage toward businesses he wishes to support. The power of the chief rabbi, then, extends well beyond his formal mandate, and it can easily trespass onto areas formally under the control of the Board of Representatives.

Since the constitution of the MT in 1814, standoffs have occurred regularly between the chief rabbi and the MT president, as each has asserted the right to determine the shape of the Jewish community in Denmark. The most public confrontation occurred in 1913, with the dismissal of Tobias Lewenstein (see chapter 1). Following that scandal, the RS tried to limit the power of rabbis, first by hiring them for five-year contracts rather than for life, and later by spelling out the limits on rabbinical authority in greater detail. These changes have not eliminated the conflicts, nor have they led to a subservient rabbinate. Indeed, since World War II, the chief rabbinate in Denmark has taken on the style of a dynasty. Marcus Melchior held the position from 1947 to 1969, hiring his son Bent as assistant rabbi in 1963. Bent Melchior assumed his father's position in 1970, then turned the office over to his protegé Bent Lexner in 1996. The only other major candidate for the position at the time was Bent's own son, Michael, who declined to leave his thriving congregation in Oslo. Both in public perception and in administrative influence, the position of the chief rabbi has steadily eclipsed the MT president's for the past half century.

One of the most florid examples of this conflict came in April 1981, when the political equivalent of a bare-knuckle brawl broke out between Bent Melchior and the MT president, Finn Rudaizky.[9] The dispute initially turned on a relatively mundane issue, the authorization of a television documentary on the Danish Jews. Melchior had arranged the program with the state television agency; it included an interview with the rabbi, footage of Passover services, and short clips of Jewish institutions like the Carolineskole and Meyers Minde. He had neglected, however, to ask permission from the RS, and when word of the program reached the board, several members expressed concern. Political events in the Middle East had reached new heights of tension in the preceding months, leading to worries about Palestinian attacks on Jewish institutions in Europe. What if terrorists used the footage to plan bombings in Copenhagen? The board therefore instructed Melchior to tell the station to remove all footage of Jewish institutions from the documentary. Melchior refused; he regarded the threat as overblown, and in any case he doubted whether the producers would agree to remove the footage. The board members then threatened to fire Melchior unless he complied, and when he still refused, they voted by a bare majority to terminate his contract.

This drastic response to a relatively petty disagreement had both personal and institutional roots. Melchior's media presence had created frictions within the MT for years; an outspoken advocate of leftist political causes, he had a zest for public appearances that some Jews regarded as self-promotion. A decade earlier, a previous MT president had tried unsuccessfully to discipline Melchior for some of his statements in television interviews. For Rudaizky, the 1981 episode marked a clear violation on Melchior's part, one that could be used to bring him to heel. At the same time, however, long-standing personal animosities contributed to the RS majority's decision. The vice-chair, Erik Guttermann, had feuded with Melchior for decades, and their mutual contempt was common knowledge in the congregation. While Rudaizky and two of his RS supporters all came from the Youth List, many in the congregation saw Guttermann as the real source of the conflict.

The termination of Melchior's contract may well have been intended as a gambit, a first step in negotiating a clearer demarcation of the rabbi's role. Its result, however, was explosive. Melchior hired a lawyer and challenged the termination as groundless and illegal. The dissenting minority on the RS voiced its agreement, and a groundswell of anger swept through much of the community. The Congress of Delegates reacted strongly, voting unanimously to demand a reconsideration by the RS. Backed into a corner, Rudaizky and his supporters refused. They defended Melchior's firing as legitimate, and they dismissed demands for their own resignations. Attempts by third parties to intervene bore little fruit; while various compromises were suggested, Melchior refused to give ground on his basic contention that he had done nothing warranting dismissal.

From a political viewpoint, Melchior's rigidity made sense. His contract required a year's notice of termination, meaning that he would remain chief rabbi until May 1982. In March 1982, DF elections were scheduled that could bring in a new Board of Representatives. With a majority of the congregation clearly supporting him, Melchior campaigned energetically for his own retention.[10] A new voluntary association was quickly organized, the Action Committee for the Keeping of Bent Melchior as Chief Rabbi. Melchior's supporters organized a mass meeting on his behalf in the Copenhagen Odd Fellows Hall, producing a petition in his favor with seven hundred signatures. His supporters secured the endorsement of the vast majority of the MT's voluntary associations, with groups from WIZO to Hakoah weighing in on his side. Melchior himself displayed a flair for the dramatic that contrasted with the clumsy conduct of Rudaizky's group. At one DF meeting in May, for example, Melchior attended despite Rudaizky's explicit request that he not do so. An article in the community journal described the ensuing scene:

> The tension reaches a new high point, when Bent Melchior asks for the floor. This the chair [Rudaizky] refuses to give him. Anger grips the audience, and tumultuous scenes break out. It is as though all these fine people have forgotten their everyday existence and are willing to mount the barricades. For their rabbi. Against their president. But he is inflexible. It is he who must lead the meeting, and nobody else. And a rabbi, who is present physically but not formally, cannot take the floor. It doesn't help matters that the delegates who initially supported Finn Rudaizky this evening now appeal to his good sense and good will. He closes the meeting and leaves, to shouts of "Shame!"
>
> Now, when the DF meeting has been ended so abruptly, people refuse to go. Tense, almost breathless, they listen to Bent Melchior's words. He describes the situation as he sees it, and regrets the obvious communications breakdown from the Representatives' side. (Yaari 1981: 10, my translation)

Over the next eight months, campaigning in the community became increasingly shrill and bitter, covering none of the parties with glory. The representatives used all means available to advance their case, including taking over the community journal and replacing its editors. They made new accusations against Melchior, suggesting financial improprieties and clandestine politicking elsewhere in Europe. Melchior's supporters, meanwhile, directed torrents of abuse at the RS majority, including threatening letters and harassment of their children in school. Some of the hostility exposed longtime rifts in the congregation. Mietek Szpirt, an RS member born in Poland, received several hate letters referring to his origins, including the following anonymous screed ultimately traced to the chairman of the Action Committee:[11]

> You are one of the most hated men in the Mosaiske Troessamfund in Copenhagen . . . you have been in this land ca. 10–12 years, and you think you have the capacity to be competent to lead or be part of leading the MT. You think—maybe because you are a doctor—that you know the Danish mentality, but you just don't, or you would have understood the consequences when you voted against B.M. You are for many years a dead man in Jewish politics. . . . A little committee—of men you don't know—is looking into whether your Danish citizenship can be taken from you, since you are not wanted here. . . . You are a miserable guest worker, whom no one has invited. But before you leave Danish soil, remember to go up to Bent Melchior and apologize. . . . Go away, and never come back, you miserable pest. OUT.[12]

The letters, when published, produced outrage in the Polish community, but the anger did not translate into support for the representatives. Indeed, Szpirt's own wife became an outspoken supporter of Melchior. Guttermann became a sort of bogeyman among liberals, who at one point mounted nighttime patrols in response to a rumor that he had imported thugs to vandalize the Jewish cemeteries. When the March elections finally rolled around, Rudaizky and his supporters endured a resounding defeat, with only Guttermann's orthodox following standing by them. The new board was solidly pro-Melchior; it rescinded the termination of his contract and stood firmly behind him for the rest of his tenure.

Melchior served for fourteen years after that election, effectively dominating politics in the congregation. When I began working with the Danish Jews in 1996, I was told by a variety of informants that the RS had become a rubber stamp for his views. He was enormously popular among liberal Jews, who almost always described him to me in glowing terms. When Melchior retired later that year, however, the political balance between rabbi and RS began to change. Bent Lexner has some important resources within the congregation; far more conservative than Melchior, he has built bridges with Guttermann and other orthodox members, while his work at the Carolineskole has given him a strong base among MT youth. At the same time, however, many of the liberal Jews who had supported Melchior so enthusiastically look at Lexner with considerable suspicion. His strict views on admission to the Carolineskole, on the conduct of ritual, and especially on conversion struck many of those I interviewed as a step backward. In 2000, liberals translated this concern into votes with the selection of Jacques Blum as the new MT president. Blum—intermarried, liberal, a sociologist with a long history as a critic of MT institutions—represents almost a polar opposite to Lexner's neoconservatism, and clashes between the two on MT policy seem inevitable. Whoever wins, the structural conflict between the chief rabbi and the Board of Representatives is likely to endure.

Conclusion

An interesting sidenote to the rabbinical battles of 1981–1982 was their effect on Erik Guttermann. Regarded by many as the mastermind of the effort to dismiss Melchior, Guttermann found his position decisively rejected by the MT majority; he lost his position as vice-chair of the RS, along with much of his political influence. He did not, however, lose his seat on the RS. Since elections to the DF use proportional representation, minority parties can still win seats, and Guttermann was re-elected along with three pro-Melchior representatives. He also remained an influential member of several important voluntary associations, where he continues to play a leading role today. While Melchior may have toppled his opponent in a political conflict, he could not—no matter how overwhelming his victory—drive him completely from the field of battle.

This pattern characterizes the organizational world of the Danish Jews generally. The structure of community institutions allows virtually all Jews to play a role if they want to and makes it almost impossible to keep any of them out. Association life caters to a broad swath of Danish Jewry, with groups representing a wide range of activities and ideological viewpoints. Jewish institutions cast a narrower net, restricting access to members of the MT; still, any member may take part in them, and one finds representatives of every subgroup in the halls of the Carolineskolen and the graves of the Jewish cemetery. The political system encourages participation by all interested parties, and it offers a chance for even marginal viewpoints or community newcomers to gain real power. The Mosaiske Troessamfund provides a single nexus through which almost any self-ascribed Jew in Denmark can find some sort of institutional involvement.

At the same time, the organizational scope of the MT makes the development of alternatives to it very difficult. I asked several of my liberal informants why they didn't start a new synagogue; since so many of them disliked the orthodoxy of the MT, why didn't they establish a Reform group themselves? The answer, in most cases, was essentially institutional. Where would they find the money for their own graveyard, if they could no longer use the MT cemetery? Where would they house their elderly, if not the MT senior homes? How could they maintain an association life, when all of the major Zionist and recreational groups had ties to the MT? The comprehensive scope of the MT makes a schism or a new congregation difficult to contemplate. It is almost always easier to work within its extremely inclusive structure than to build something outside of it.

As a result, the MT creates a peculiar and powerful sort of community among Danish Jews. The people I interviewed in Copenhagen came from a wide variety of backgrounds; their lives, their occupations, their views of the world and of Judaism differed radically. They had in common, however, a single set of institutions through which they articulated their understandings of their identities and their heritage. Whatever it was they thought of Judaism or Jewry, they had a single institutional context within which they could express it. This context, moreover, brought them together socially. With a few exceptions, they tended to know the other Jews in the city, or at least to know of their families, because of their common participation in organizations and institutions—they knew each other from school, or from association boards, or from debates in the Congress of Delegates. It is hard to define what a Jew is, in Denmark or anywhere else, and it is even harder to define the outside limits of the Jewish community. At least in Copenhagen, however, it is possible to define the center. Much though they disagree about it, much though they fight about it, much though some of them may even detest it, it is the organizational structure of the Mosaiske Troessamfund that constitutes the Danish Jews as a group.

The Larger World

Relations with the Jewish Community Outside Denmark

I live in the state of Indiana, in a small city about an hour northwest of Indianapolis. To many people on the coasts of the United States, I might as well live on Mars. Residents of New York and California tend to regard the Midwest as a featureless wasteland, and I have spent a good deal of time at anthropology conferences explaining to other Americans exactly where Indiana is. I was surprised, then, when my first informant among the Danish Jews told me he knew all about the place. He had a degree in pharmacy from Purdue University and was able to describe the building where I work in considerable detail. Another informant had relatives in nearby Lafayette, whom she asked me to look up when I got back to the United States. Others referred me to relatives elsewhere in the Midwest, and still others told me about traveling through my area on trips across the United States. It became obvious very early in my fieldwork that America was familiar territory to almost all of the people I was interviewing, a place they associated with family, education, work, and tourism. They knew my home far better than I knew theirs, and indeed they had sometimes seen more of it than I had myself.

It wasn't just America, either. The world of the Danish Jews extends well past the perimeters of Copenhagen, to Europe and the Middle East and sometimes beyond. All of them have relatives in other countries, almost all of them have traveled abroad, and a surprising number have lived abroad for years at a time. In this the Jews of Denmark resemble other European Jews and share in a basic element of the Jewish experience. Since the destruction of Israel in the second century, Jewishness has been an identity that transcends national borders; Jews have been dispersed among a variety of countries, and their links to coreligionists in distant places have often had more social significance than their relationships with local neighbors. Jewish history is almost uniquely international, involving repeated displacement to almost every corner of the globe over a period of two thousand years. As a result, it is impossible to describe Jewishness in any particular spot without reference to the worldwide whole, the international network of kinship and cultural links through which all Jews must, at least in part, understand themselves.

While Jewishness always has an international aspect, the shape and influence of that internationalism vary from place to place. Different Jewish groups encounter the larger world in different ways, and they use different local contexts to interpret it. The meaning of Israel to a seventeenth-century Polish shtetl, for example, differed radically from its meaning to a late twentieth-century American Reform congregation (Furman 1987; Zborowski and Herzog 1952). Residents of the shtetl met Israel through the lens of religion and folklore, as the site of a revered mythic tradition and a symbol of a promised messianic redemption; the Americans met it through travel and through the press, like any other contemporary state, and most of them believed neither in the myths nor in the Messiah. To understand the role of Jewish internationalism in any particular community, then, we must first understand the specific ways that its members make contact with the larger world and the cultural categories through which they interpret it.

This chapter takes such an approach to Jewish Copenhagen. It discusses the kinds of contacts—kinship ties, population transfers, cultural influences, and others—that Danish Jews have with the larger world. It then considers the effects that these links have on the nature of Jewish community life in Denmark. International contacts promote specific kinds of social change in the Mosaiske Troessamfund, including the demographic weakening of the orthodox wing. They also create a number of cultural patterns, politicizing Jewish identity while creating distinctive models of Jewish difference. This chapter explores some of these patterns and processes, and it suggests some implications for the nature of community in Jewish Copenhagen.

INTERNATIONAL CONNECTIONS AMONG COPENHAGEN JEWS

For most Copenhagen Jews, a knowledge of the world beyond Denmark begins in early childhood, intertwined with family history and personal relationships. All Jews trace their family roots beyond the country, and many have immigrants as parents or grandparents; the family stories told to children recall distant places and cultures, as well as the historical events that brought the ancestors to Denmark. In most cases, Jews also have relatives in other countries whom they meet and visit as children. As they grow older, they encounter a steady flow of information about Israel in Jewish kindergartens and the Jewish school. All of the Carolineskole students visit Israel in their final year, and many of them spend time there earlier with their families. As a result, even living in a cosmopolitan setting like Copenhagen, Jews show a remarkable ease with both the notion and the process of moving between cultures. I met small children during my research who spoke of a trip to Israel as if it were a drive to the supermarket; they knew more than I did about passports, security, and air travel, and they had strong opinions about the different kinds of Middle Eastern food. When I interviewed adults, many of them discussed moving to other countries in similarly matter-of-fact tones. Most Danes, of course, travel abroad repeatedly during their lifetimes, but they do so primarily as tourists. The Jewish openness to complete relocation, the sense that other places offer not only an interesting diversion but a viable potential home, was one of the most striking distinguishing features of my informants as a group.

Foreign connections among contemporary Jewish Danes tend to concentrate on a few particular world areas—Israel, the United States, England, and Scandinavia. Each of these places has a distinctive significance to Danish Jews, and each tends to involve a different kind of contact. Jews know about the rest of the world, of course, and like most Danes they tend to have seen some of it. Beyond these four areas, however, their interest in the outside world differs little from that of non-Jews, and their contacts tend to be through standard business or tourist interactions. While many contemporary Jews have family roots in eastern Europe, for example, few have contacts there today; I never met one who seriously contemplated relocating there. The key areas of Jewish interest involve ongoing exchanges of people and cultural information, which vary according to the specific relationship between those areas and the Danish community.

By far, the most important of these areas is Israel. The Copenhagen Jews have a range of ties with Israel, ranging from personal networks to formal institutional affiliations. Some of the formal ties run directly through the MT; when an Israeli dignitary visits the country, for example, the MT president and rabbi act as the official representatives of the Danish Jews. Other institutional relationships involve voluntary associations like Hakoah and

WIZO, which function as the Danish branches of worldwide organizations headquartered in Israel. On an individual level, most Danish Jews have some direct personal connection to the country. A significant number were born there; about five hundred Israeli émigrés live in Denmark, a number that has been slowly climbing for a decade. Even more have gone the other way. Many Danish Jews have emigrated to Israel since 1945, so that almost every Jew I interviewed had a child, a sibling, or at the least a cousin living there. The overwhelming majority visit Israel as tourists, and many have lived there for extended periods. As a result, Israel has real sense of social closeness for Danish Jews. When they travel there, they arrive not as strangers but as family, people with roots in the society and a personal interest in its welfare.

Beyond actual travel and emigration, Israel has a powerful ideological significance for Copenhagen Jewry. Like most Diaspora Jews, they tend to regard Israel as the symbolic center of the Jewish world. Its history and religion lie at the heart of Jewish culture, and its current inhabitants represent the first autonomous Jewish state in two millennia. The character and survival of that state therefore constitute vital concerns for most Jews in Copenhagen, and they follow its political and military affairs with interest. Many Danish Jews know more about the Israeli parliament than the Danish one; in my interviews, I heard far more discussion of Bibi Netanyahu and Ehud Barak than about the current ministers of Denmark. Beyond politics, the culture of Israel represents an important model of Jewish authenticity for the Copenhagen community. With its vibrant and tumultuous Jewish life, Israel seems for many to represent a truer form of Jewish ethnicity and religiosity than their own. They look to Israeli styles of dress, grooming, speech, and affect as indexes of the Jewishness of their own behavior. They also pay considerable attention to Israeli culture and history, particularly in their publications. Danish Jews tend to regard their own community as something peripheral, a little outpost of Jewishness far removed from the real centers. The key center, and the primary object of their concern as Jews, is Israel.

Another center, not as important but certainly much closer, is England. With a Jewish population of about 280,000, England offers a source of abundant Jewish life easily accessible across the North Sea. It also offers an accessible culture. Most Copenhageners speak very good English, and the climate and affective style of Great Britain are familiar. As a result, many Copenhagen Jews have visited or spent time in Jewish enclaves like those in Golders Green and Birmingham. For the orthodox, especially, these areas offer a Jewish environment that Denmark does not—real Jewish neighborhoods, with kosher supermarkets, Jewish restaurants, and a variety of active synagogues. They also offer something very difficult to come by in Denmark: marriageable Orthodox Jews. A number of my orthodox informants had found their spouses in England, and others were encouraging their unmarried children to study there for that reason. England does not have the cultural influence of Israel in Denmark, and it has no more institutional significance than any other European country, but it forms an important social resource for the community's orthodox wing.

Something similar could be said of Sweden and Norway, Denmark's nearest neighbors in cultural terms, both of which have small Jewish populations. Norway has about a thousand Jews. Most live in the capital city of Oslo, where Michael Melchior presides over a single Orthodox congregation. A very small congregation also exists in the far northern city of Trondheim. Sweden's Jewish population is much larger and more widely distributed. Of its 17,000 members, about 5,000 live in Stockholm, 2,000 in Malmö, and about 1,800 in Göteborg; other small congregations are scattered about the country. Sweden also has a greater religious diversity, with a Conservative synagogue in Oslo and two in Malmö.

The histories of Denmark, Sweden, and Norway have been intertwined for more than a millennium, and at various times they have formed a single national entity. Their languages are mutually intelligible, and their cultures have an unmistakable similarity in symbolism and aesthetic style. Moving among the three countries is therefore easy and relatively common, for Jews as much as anyone else. Many of my informants had worked or lived in Sweden or Norway for a time, and a number had family connections there. Some even commuted daily to offices in southern Sweden. Swedes and Norwegians in Copenhagen, likewise, can take active roles in the MT. The Peace List, for example, was founded by a Swedish psychology professor. Jewish institutions in the three countries have strong relationships. Beyond the regular contacts among the region's rabbis, a number of voluntary associations carry on joint projects. These relationships can be very helpful at times—between 1943 and 1945, for example, Swedish Jews helped some Danish refugees find food and jobs. More recently, Denmark has supplied Sweden with kosher meat since the banning of ritual slaughter there.

To some extent, ties to the United States among Danish Jews resemble those among Danes generally. Denmark has an unusually close relationship with the United States, in large part because so many Danes emigrated there during the nineteenth and early twentieth centuries. Denmark celebrates the Fourth of July, for example, with a lively festival in the village of Rebild that features a prominent American as speaker. Large numbers of Danes travel to America as tourists, and it is all but impossible for an American not to be moved by the genuine affection for the United States that one meets while living in Denmark. In my work with Danish Jews, I found a similar degree of closeness, although seldom the same sort of affection. Like other Danes, most Jews have relatives in America; the Eastern European Jews who arrived in Denmark during the twentieth century were a tiny part of a massive wave of emigration, most of which crossed the Atlantic. Like other Danes, Jews often travel to America as tourists, and many have worked or studied there as well. They also receive a substantial number of American Jewish tourists, many of whom stop in for services at the Krystalgade synagogue in the summer. Despite its distance, America constitutes a regular source of social contact and population exchange.

Just as important, Americans have a significant ideological role in Jewish Copenhagen. American Jews, unlike Danish ones, constitute a visible and influential force in intellectual and expressive culture; American Jewish filmmakers, artists, and writers have created a body of work exploring the meaning of the Jewish experience that has no real parallel in Scandinavia. Accordingly, Danish Jews often draw on American culture as they work out ideas about the nature of Jewish culture and identity. My informants often referred to books by writers like Saul Bellow, Philip Roth, and Chaim Potok, whose novels about Jewish life in a Christian culture had resonance for them. American Jewish films were very popular as well, particularly those directed by Woody Allen. Danish Jews also look to the United States for models of religious engagement. Those in favor of congregational reforms often used the American Reform movement to explain their goals to me, while the orthodox often expressed admiration for the Hasidic communities of New York. With the diversity of its Jewish communities, and the richness of its Jewish expressive culture, America offers a variety of models through which Danish Jews can understand their own experiences.

Israel, England, Scandinavia, and the United States dominate international interactions among Copenhagen Jews. Ties to other countries are much less extensive and have much less impact on the life of the community. They do exist; Jews travel and do business in Europe like other Danes, and some Jewish organizations take part in pan-European activities. Bent Melchior, for example, was president of the European B'nai B'rith for much of the 1990s. Europe also provides a variety of Jewish resources that some Danes take advantage of, ranging from cultural sites like the Prague synagogue to special kosher foods imported from France. Danish Jews have engaged in relief efforts for troubled Jewish communities in various parts of the world; Russian Jews have been a subject of particular concern. Such activities, however, are sporadic and largely individual. While of great importance to some individual Danish Jews, they are not, like Israel or America, basic elements in the shape of Jewish experience in Denmark.

IMPACTS OF EXTERNAL CONNECTIONS

Connections to the larger world have important effects on Jewish life in Denmark, both for individuals and for institutions. For individuals, foreign contacts shape understandings of what Jewishness means, placing their identities in a context far larger than their local experience. For institutions, the external world provides both resources and dangers, icons that bind their members together as well as issues that tear them apart. To get a sense of these impacts, I would like to focus on four ways that external connections affect the Copenhagen Jewish community as a whole: the extent to which they provide a focus for community activity; their role in defining Jewish authenticity; the population exchanges that they promote; and their role in politicizing Jewish identity and activities.

Focusing Community Activity:
Israel and the transcendence of community divisions

One summer evening in 2000, I visited an MT official at his home in central Copenhagen. I arrived around ten o'clock, at his request, and found him just getting back from a meeting down in Ny Kongensgade. He had spent several hours there, and he looked exhausted; the meeting had been a tough negotiating session, the culmination of months of preparation, a showdown between three groups whose positions had seemed hardened and irreconcilable. The subject? A proposal to open a kosher café in the basement of the MT offices. After years of effort, he told me, a bargain had been struck, and he thought the café could get approval within a few months. He was happy about that, but his frustration with the process was almost palpable. "That's the problem with this community," he said. "There are so many little factions, and they fight so much about everything, that you can never get anything done."

This problem presents a real challenge to institutional activity in Jewish Copenhagen. With so many divisions within its membership, common action becomes difficult; disagreements can paralyze even so innocuous a project as kosher café, and they can make more ambitious programs seem impossible. In such a setting, how can group activities bring together broad sections of the Jewish community? The answer, in many cases, involves the world outside of Denmark. Many of the divisions within the MT derive from local history—family rivalries, immigrant experiences, patterns of institutional control, and so on. Standing outside of this history, the larger world can transcend those divisions. Most of the events and associations that draw large parts of the community together base their appeal on an external foundation, some cause or organizing figure that stands apart from the local world. The international dimension of Jewishness in effect provides a means of escape from the organizational paralysis of a fractious local community.

In most cases, such unifying activities involve the state of Israel. Orthodox or liberal, Viking or immigrant, insider or outsider, almost all Jews regard Israel's survival as an important concern, and they want to maintain their family contacts with the country. As long as it steers clear of Israeli politics, therefore, an organization or activity that aims to support Israel can count on broad Jewish support. The most encompassing voluntary organizations in the MT, for example, are those that support Israel—WIZO, the United Jewish Appeal, the Danish Zionist Association, and others. These groups deliberately avoid mixing in issues that could divide their Danish supporters; they do not favor one or another Israeli political party, and they avoid discussions of divisive issues like the status of Israel's occupied territories. Where they support specific Israeli institutions, they tend to choose ones that have universal approval, such as the Institute for the Blind. One Jewish man explained the success of WIZO to me as a result of the group's focus on daycare centers. "They have it easy," he said. "Who's going to object to sending money to a children's home?" Whether such groups do much for Israel is an open question; my informants regularly dismissed the economic importance of the Danish Zionist groups, saying that Israel didn't need their money and that in any case many small neighborhoods in New York probably contributed more each year than the entire Copenhagen community. Their importance to Jewish Denmark, however, as a means of transcending local divisions, is beyond doubt.

The disconnection of outsiders from local social cleavages can sometimes have implications for the status of immigrants. The arrival of the Polish Jewish refugees in 1970, for example, provided a unifying cause for the existing Copenhagen community (Roth 1999). Not only did these new Jews stand outside of local disagreements, their expulsion evoked memories of the Danish Jews' exile during the war. People from all branches of the MT joined in the support of the Poles, first in lobbying for their admission to Denmark and later in helping to get them settled. Jewish journals at the time exulted in the common feeling that the Poles' plight had occasioned. A related phenomenon followed the arrival of two Chabad missionaries from New York in 1996. As noted in chapter 3, the missionaries have organized a variety of Jewish holiday festivals, each of which has attracted an enthusiastic turnout from all subgroups of the MT. Their lack of implication in local rivalries has enabled the missionaries to draw together Jews who could not cooperate in locally organized events. Such outsider status, of course, has a limited lifespan. Over time, newcomers become part of the local scene, and they become parties to the conflicts that they had once enabled others to transcend. The unity that surrounded the Polish immigration dissipated as the Poles became established in Denmark; old divisions reasserted themselves, along with new oppositions between the immigrants and the Viking Jews. So far, the Chabad leaders have avoided being drawn into the political life of the MT. As their movement grows, however, they will inevitably form alliances and create new power centers, and they may find their audience narrowing considerably.

Modeling Jewish Identity and Practice

The larger world provides the Copenhagen Jews not only with causes but also with models—models of what Jews should look like, how they should behave, and how they should practice their religion. There are local models too, of course, the family and social connections that form individuals' primary understandings of Jewishness. International models, however, make up an important alternative, particularly for Jews seeking an especially distinctive identity. On the whole, Danish Jews tend to look and behave much like other Danes; while a careful observer can pick out distinguishing features, such details do not add up to a characteristic physical or personality type. Those seeking a vision of "authentic" Jewishness, an image of Jews as something different and identifiable, often look elsewhere. As youths try to solidify their inchoate ideas of self, or as nonpracticing adults try to conceptualize an ethnicity that has few concrete manifestations in their lives, they often turn to the striking models of Jewish difference in the international Jewish community.

Such models come overwhelmingly from Israel. Israel's prestige in the Jewish world, its dramatic historical and political setting, and its drastic contrast with the bourgeois world of Northern Europe combine to make it a favorite model of authentic Jewishness among Copenhagen Jews. Jewish teenagers I met spoke glowingly of Israel as a place where "real" Jews were, and most of them thought actively about moving there. Some of them imitated Israeli styles of dress and grooming, with small knitted skullcaps especially popular among boys. Adults, too, tended to think of Israelis as more genuinely Jewish than themselves. The imagery is everywhere. The walls of Jewish institutions teem with posters of Israel, depicting scenes of smiling, dark-featured Jews and dramatic Jerusalem scenery. Objects from Israel stand in prominent places in both institutions and homes; I seldom saw a Jewish home without at least one Israeli souvenir displayed on a shelf, and most had more. The Carolineskole curriculum focuses heavily on Israel, teaching students about its history and culture and encouraging them to emigrate there. Jewish kindergartens read children's books about Israel, Jewish magazines feature articles about Israel, Jewish websites and publications use the Israeli flag as an emblem. Even for Jews who never go there, Israel plays a powerful role in defining what it means to be Jewish in Copenhagen.

This model has institutional effects as well. In the perennial battle between the orthodox and liberals over MT ritual practice, for example, Israeli religiosity forms an important weapon in the orthodox arsenal. The state religion in Israel is Orthodox Judaism; alternative denominations like Reform and Conservatism have no standing there, and the ritual actions of their rabbis have no official force. This policy helps to legitimize an Orthodox approach in Copenhagen, associating alternatives with departures from authentic Jewish practice. It also creates administrative obstacles for non-Orthodox practice. A man who converts to Judaism in a Reform congregation, for example, may not have that conversion officially recognized if he moves to Israel. This lack of recognition may affect his ability to obtain Israeli citizenship, and it may have consequences for his marriage and the civil status of his children. When liberals suggest reforms in MT ritual practice—for example, allowing men and women to sit together in services or allowing women to read from the Torah—conservatives respond that such changes would depart from Israeli practice, and hence would threaten the validity of Copenhagen conversions. The warning is not idle—Bent Melchior at one point lost the approval of the Israeli rabbinate because of his lenient conversion policy (Anonymous 1981). Israeli notions of authentic

Jewish practice, in other words, carry both symbolic and administrative weight in Copenhagen, helping to reinforce a particular approach to ritual.

It would be wrong, however, to suggest that Israel alone has an influence on Copenhagen notions of Jewish authenticity. Other Jewish centers offer competing visions of Jewish ethnicity and religiosity, and some MT members draw on them as they consider changes in the community. Partisans of ritual reform, for example, often use the American Conservative movement as a model; a number of my informants said that they had visited Conservative synagogues while traveling in the United States and that the experience had convinced them that reforms in Copenhagen could entice Jews back into the synagogue. Orthodox partisans, on the other hand, often mentioned the Hasidic communities of New York as evidence that a strictly observant lifestyle could thrive in a modern social setting. The Chabad missionaries have given this influence a tangible form, presenting a radically different understanding of Jewish religiosity than those traditionally dominant in Copenhagen. While their following is still quite marginal, they are widely known in the Jewish community, and many of my informants expressed admiration for their rigorous and unabashed religious observance. Though they have had little success in converting Copenhageners to Hasidism, these Americans have become icons of Jewish authenticity for large sectors of the congregation.

Population Exchanges

Not only ideas flow between Denmark and the larger Jewish world; people do as well. For centuries, international connections have both brought Jews to Denmark and taken them away. All Danish Jews, of course, are descendents of immigrants, and the majority have ancestors who immigrated within the past hundred years. The immigration continues today, a small but steady trickle from the Middle East and occasional arrivals from elsewhere. At the same time, virtually all Danish Jews have a connection with emigration. A substantial number have moved to Israel since 1948, and smaller contingents have gone to England or the United States. Even for those who stay, emigration to Israel remains an option that they have at least considered. Though the last major wave of immigration lies a generation away, then, a steady pattern of population exchange still goes on in Copenhagen, slowly and quietly stirring the demographic stew of the Jewish community.

What makes this quiet process so important is that immigration and emigration stem from very different factors. Jews who emigrate do so, for the most part, because of a strong commitment to Jewish identity. They move to Israel or England, where they can find a vibrant Jewish community and easy access to the necessities of Jewish practice. This trend has occurred especially among the orthodox, and it has thinned the ranks of Machsike Hadas drastically (see chapter 3). It also affects the composition of the community's younger cohort; the Carolineskole strongly encourages emigration to Israel (*aliyah*), and its most religious students often take up the call. Even outside the orthodox institutions, therefore, emigration tends to draw off the most religiously engaged segment of the Jewish community.

If people tend to leave because of a commitment to Jewish life, they tend to come because of an attraction to Danish culture. In many cases, that motivation includes an attraction to a specific Dane: many of the Israeli immigrants I met had moved in order to be with a Danish spouse or lover. The others came for a wide variety of reasons, ranging from work to study to an interest in Danish art. For all of them, the relationships or interests that had brought them to Copenhagen outweighed the loss of Jewish fellowship that the journey involved. Those who were religiously observant often went to Machsike Hadas for services. More often, however, recent immigrants have very little to do with Jewish institutions. They do not come to Denmark out of an interest in the Jews there, and that interest seldom blossoms after their arrival.

As a result, population exchanges tend to deplete the orthodox wing of the MT and to add to the secularized Jewish population. Over time, this pattern has brought the orthodox faction to the brink of collapse. Members of the group describe this prospect with some sadness but also a degree of pride. Machsike Hadas has declined, after all, not because members have lost their faith but because they have prized it; its young people have not forsaken Judaism; they have moved to its homeland. Its members remain close to one another in new homes in Jerusalem or Golders Green. The decline of Machsike Hadas is a loss less for its members than for the Jewish community as a whole, which over the decades has seen the ranks of its most committed members slowly eroding.

Politicization of Jewish Life in Denmark

When Danish Jews talked to me about Israel, they talked mainly about its heritage and culture. Israel is also, however, a state, a political entity set in one of the world's most tumultuous and fragmented geographic regions. From its inception, Israel has been embroiled in explosive political disputes that have sparked strong opinions around the world. The ties that bind the Danish Jews to Israel therefore also carry a political dimension, a tendency to associate Jews with political turmoil in the Middle East. This pattern characterizes both the MT itself, where differences on Israel can lead to divisions in the congregation, and the perception of Jews by the wider Danish society. Internally and externally, Judaism's international dimension acts to politicize the local community.

The external dimension affects almost all Jews, who can find themselves drawn into debates over Israeli policy in workplaces and social groups. As one man put it to me, "I don't happen to know all that much about Israel; I don't follow politics very much. But everyone at work assumes that I know all about it, and that I'm a spokesman for whatever Israel is doing. They'll come up to me and say, 'How can you do this to the Palestinians?' And what shall I say?" This position has become increasingly uncomfortable in recent years, as the image of Israel has turned more and more negative in the national press. Since the invasion of Lebanon in 1982, and especially since the beginning of the second Intifadah in 2001, press coverage of Israel has focused largely on accusations of human-rights violations. The Palestinians, once a synonym for terrorism in Denmark, have become objects of widespread sympathy, especially on the left.[1] As a result, Danish Jews find themselves called upon to defend what many Danes see as a repressive government. In the summer of 2001, for example, when Israel appointed former Shin Bet director Carmi Gillon as ambassador to Denmark, several Danish human-rights groups demanded the arrest of Gillon, who had recently endorsed a form of interrogation widely regarded as torture. The affair drew international attention, creating a dilemma for Copenhagen Jews: to denounce Israel's representative in Denmark or to defend what their neighbors saw as torture. The nation's most prominent Jewish politician, Transportation Minister Arne Melchior, defended Gillon publicly; the resulting firestorm of criticism led to his expulsion from the Center-Democrat Party which he had founded twenty-eight years earlier. While the MT assiduously avoided any public comment on the issue, both the institution and its members endured a difficult few months.

This association of Danish Jews with Israel comes not only from Danes but from the disputing parties themselves. Israeli officials expect Danish Jews to support them; when a Jewish magazine in Copenhagen recently published an article critical of Israel, for example, the Israeli ambassador is said to have prevailed upon the MT leadership to have the editor dismissed. Palestinians in Denmark, meanwhile, sometimes treat Danish Jews as proxies for Israel, attacking both institutions and individuals. The most violent episodes occurred in 1985, when a group of Palestinian terrorists bombed the Krystalgade synagogue and a kosher food processor. More typically, individual Palestinians have insulted or threatened individual Jews during periods of high tension in the Middle East. One woman, who operates a Jewish café, told me of being threatened by Palestinian youths when she first opened the store; she called the police, who chased the men but never caught them. Minor confrontations occur occasionally outside Jewish institutions, and one orthodox Jew told me of being knocked down and insulted by a Palestinian man. The most organized recent incident occurred in the late 1990s, when a group of Palestinian youths attacked a Hakoah soccer team as it arrived for a game. Such events do not occur often, and they have yet to produce a serious injury. In the Copenhagen context, though, where violence and anti-Semitism are normally very rare, they constitute a disturbing and unsettling intrusion. They also complicate relations between Muslims and Jews, who otherwise have significant common interests in Danish politics.[2]

Within the community, attitudes toward Israel can produce serious political rifts. The rifts do not involve actual oppositions to Israel, which is virtually unheard of. Rather, Jews differ on how much criticism is consistent with support for the state. Do Danish Jews have the right to criticize the policies of the Israeli government? And if they do, should they voice that criticism outside of the Jewish community? Some of my informants answered both questions with a strong negative. As one put it, "I don't live in Israel, and I don't have to take any of the risks and dangers associated with living there. I think that the ones who make those sacrifices are the ones who should have a voice in Israeli policy. Those of us in the periphery should just show support." The MT generally takes this position, as do such mainstream Zionist organizations as the DDI and the DZF. Other Jews disagree, however, and at times their views can produce real cleavages in the congregation. In the aftermath of the Israeli invasion of Lebanon in 1982, for example, angry Jewish leftists organized a voluntary organization called New Outlook (see chapter 3). Associated with the Peace Now movement in Israel, the group called publicly for a Jewish critique of Israeli policies. Their actions enraged many members of the congregation, including the rabbi, and some were one threatened with expulsion from other Jewish groups. The same crisis also gave birth to the Peace List (*Fred-*

slisten), which gained a number of DF seats with its call for a change in Israeli policy. These movements waned along with the events that spawned them, but later crises would produce similar oppositions. As long as the Danish Jews remain attached to Israel, the disputes that divide Israelis will produce corresponding divisions in Copenhagen.

Conclusion: Jewishness as an International Identity and the Future of Copenhagen Jewry

The international dimension of Jewish identity shapes the Jewish experience in Copenhagen in two distinct ways. On a group level, it exercises a powerful influence on the social dynamics of the Jewish community. The group's close links to Israel particularly shape the demographic structure and political alignments of the MT; Israel provides both a point of unity and a point of division within the community, and it drains population from some parts of the community even as it adds it to others. One cannot describe the workings of MT politics, education, organization life, or religious observance without reference to their international framework. On an individual level, the larger Jewish world informs personal understandings of Jewish identity, ethnicity, and religious observance. As Danish Jews formulate their ideas of what it is to be Jewish, they look not only at their neighbors and relatives but also at foreign models ranging from Israeli Sabras to New York Hasids, from Yitzhak Rabin to Menachem Schneerson to Woody Allen. The Danish Jewish self, like the Danish Jewish world, is inseparable from the international world that defines it.

This influence reinforces the difficulties involved in drawing any neat boundaries around the Copenhagen Jewish community. Modern transportation and communication technologies have eliminated many of the barriers that physical distance once placed around the group; moving abroad no longer means losing touch with one's family, losing one's influence in the MT, or even giving up one's job in Denmark. One of my informants commuted between Israel and Copenhagen for over thirty years, serving in the MT leadership and participating in voluntary organizations all the while. At the same time, mass communications have brought the outside world to the heart of the community, giving people in Jerusalem and New York a direct influence on Danish Jewish identity. It would be inaccurate, I think, though some of my informants disagree, to say that a distinctive Copenhagen Jewish community no longer exists. Clearly, however, contemporary technologies have made that community harder than ever to define, as they have blurred the border between local and international Jewry.

Opinions differ on how these contacts will affect the future of the Danish Jews. When I asked my informants about it, some of them stressed the invigorating effects of immigration on the MT. In a country like Denmark, they said, infusions from the international world are a necessary corrective to the culture's powerful assimilative pressures. Both at the opening of the twentieth century and in 1969–1971, Eastern European immigrants provided a rush of new blood in a demographically shrinking community; were it not for this sort of exchange with the outside world, Danish Jewry might long since have disappeared. On the other hand, the rising prominence of outside forces since 1948 has diminished the importance of local institutions. The MT no longer encompasses the Danish Jewish world, and its leaders no longer set the standard for Jewish identity and practice. It is not the MT rabbi but the Jerusalem rabbinate that determines correct religious observance; Jews look not to the MT president but to Israelis and Americans for their models of Jewish authenticity. Perhaps as a result, the prestige and power of MT leaders has declined sharply. Powerful men like M. L. Nathanson and C. B. Henriques no longer seek the MT presidency, nor could they exercise strong authority if they did. While the international Jewish community has sometimes made the MT more vital, in recent years it has pushed the MT toward irrelevance.

It would be easy to see these events as sequential, to see the revitalizing effects as part of the past and the devitalizing ones as the trend of the future. Many of my informants, indeed, do just that. The sort of refugees who invigorated the MT in 1900 or 1970, they tell me, now go to Israel; Denmark will not see a new infusion of committed Jews again, merely an ongoing trickle of secularized Israeli expatriates and the occasional American businessman. Over time, Israel will succeed in draining away Denmark's committed Jews and turning the MT into a local ethnic franchise. They may be right. On the other hand, they may be optimistic. The current structure of the international Jewish world depends on a perilously unstable political base—it assumes the continued viability of Israel as a Jewish homeland and the continued freedom of Jews to travel there and back. Should that base fail, should Israel's survival or Jewish civil rights be significantly imperiled, the role of places like Denmark may change. The same things that attracted Jews to Denmark in 1900 or 1970—a liberal state, a culture receptive to Jews, an economic system within which Jews can prosper—may draw them there again. The current relationship between Copenhagen Jewry and world Jewry reflects a time of unparalleled freedom and prosperity for Jews

6

The Danish World

Jewish Difference in Danish Culture

The main Jewish cemetery in Copenhagen announces its identity with a discreet marble plaque by the gate. A visitor can easily miss it, as I did on my first visit there in 1996; my eyes were drawn instead to the chapel, a small brick building about fifty yards away down a broad gravel walkway. The walk is lined with tall evergreens, and in between them small sidepaths lead into the quiet rows of graves. I spent a peaceful hour ambling through the grid of gravel pathways, under a clear July sun, and didn't notice the plaque until I was on my way out. It didn't matter, though; even if I hadn't seen it, I could never have doubted where I had been. Two features distinguished the cemetery immediately, making it impossible to confuse with any ordinary Danish graveyard. One was the Jewish symbolism. Many of the gravestones had both Hebrew and Danish inscriptions, and Stars of David appeared where ordinary Danish stones would have had crosses. Holocaust imagery was there as well, with granite memorial markers commemorating Danish and Polish victims of the Nazi genocide.

The other clue, just as unmistakable, was the abundance of weeds. Most Danish cemeteries are immaculate places, their walkways and graveled plots raked and weeded assiduously. Low, neatly trimmed hedgerows mark off the different gravesites, and carefully tended plantings ornament the stones. In the Jewish cemetery, by contrast, frequent patches of green peeked through the gravel of the paths, and clumps of grass and broadleaves sprouted from around many of the headstones. In several plots I saw tall dandelions, their seeds blown away—the first I had found in

Copenhagen that summer. The hedgerows were often unkempt and over-grown; even the lawn surrounding the large Holocaust monument was weedy and uneven, with tiny saplings sprouting amidst its ragged herba-ceous border.

This seediness did not reflect any lack of reverence for the dead—quite the opposite. It stemmed from the difficulty of combining a Danish burial aesthetic with the sanctity of the gravesite traditional in Judaism. Danish Lutheran cemeteries owe their beauty in large part to their turnover. Once buried, an individual holds a gravesite for only a few decades, after which the cemetery clears the plot and reuses it. Occupants of active plots there-fore usually have close living relatives, who can undertake the intensive gardening necessary for a presentable gravesite. Jews, on the other hand, maintain graves permanently, leaving them intact even after their occu-pants have been forgotten by the living. The population of the Jewish cemetery has consequently grown inexorably over the years, even as the population of the Jewish Community has gradually fallen. The Danish style of burial, with its endless weeding and raking of the gravesite, has become increasingly difficult to maintain. The ragged walkways and weedy plots are not evidence of neglect but of strain, of the impossibility of seamlessly joining Jewish and Danish tradition. For the dead, the tension between Danish and Jewish identity manifests itself in dandelions.

The cemetery embodies an ongoing difficulty for Danish Jewry, the strain of attempting to bridge two very different cultural systems. As they make sense of themselves, as they present themselves to others, as they con-stitute their community, Jews must constantly refer to both Danish and Jewish ideological models and try insofar as possible to satisfy each of them. Doing so completely is impossible; the two systems have deep and separate roots, and they enjoin different kinds of attitude and behavior on their members. They call not only for different burial customs and food-ways, but also for different family relationships, affective styles, and notions of freedom and duty. To be a Copenhagen Jew is therefore to face an on-going conflict, a strain between two essential elements of the self that imply two different universes of meaning. It is a conflict that can be man-aged, and even settled for a time, but that never really disappears.

This problem exists to some extent for all Diaspora Jews, and indeed for diaspora populations generally, as local and ethnic cultures offer diverging models of identity and behavior. Since the expulsion of the Jews from Is-rael in 135 C.E., the task of relating Jewishness to a larger and very differ-ent host culture has been an essential element of the Jewish experience.

Some authors have contended that it gives Jews a characteristic intellectual standpoint and a characteristic personality structure (e.g., Boyarin 1996; Cuddihy 1987; Finkielkraut 1994; Gilman 1986). It poses a particular challenge to Western European Jews, for whom national identities are usually as central to individual self-perception as is Jewishness (cf. Webber 1994a; Webber 1997). They face not only the familiar Jewish task of living amidst "the other," but also a deeper one of unifying two opposed dimensions of self. The nature of this task varies from place to place, as local settings impose specific understandings of Jews, religious affiliation, group identity, and difference in general. It also varies over time, as the changing fortunes of particular Jewish communities reshape their relationships with neighbors. While all Jews must reckon with the disjuncture between local and Jewish identities, the meaning and consequences of that disjuncture take different forms in every location.

This chapter explores that process in Denmark—the ways that Jews are understood in Danish culture, the ways that Jews themselves understand Jewish difference, and the concrete forms in which the interaction of these two identities manifests itself. Two somewhat contradictory features of the Danish context have particular importance in this interaction. One is a genuine lack of anti-Semitism; the antipathy to Jews that permeates so much of European culture is largely absent in Denmark, and indeed the society at large takes a distinctly positive view of Jewish difference. At the same time, however, Danish culture tends to reject difference on a more general level, valuing national homogeneity and stigmatizing individuals who stick out. For Danish Jews, these patterns create a real difficulty. Jews see themselves as different from other Danes, and both Jewish tradition and Danish culture regard that difference as valuable. Yet Jews also share the Danish dread of difference, the pervasive concern to be an undifferentiated part of the larger group. To be Danish and Jewish therefore involves an ongoing conflict, one that makes the construction of the Jewish self a difficult and often painful project. This chapter looks at some of the conditions under which that project takes place, and their implications for the nature of Jewish community and identity in a place like Copenhagen.

Understandings of Group and Difference in Danish Culture

A recurrent theme in both literary and social scientific analyses of Danish culture is the importance of belonging, of being a member of the group

(Gullestad 1989; Gullestad 1994; Jokinen 1994; Knudsen 1996). Danes tend to draw strong distinctions between inside and outside, both physically and socially; the inside has associations of warmth, equality, and commonality, while the outside is associated with coldness and social distance. This contrast permeates Danish culture, manifesting itself in everything from design to folklore to the use of personal space. Danish rural folklore, for example, populates the interior of the farm with benign gnomes (*nisser*), while ravening trolls and other deadly spirits lurk in the woods outside. At Danish social gatherings, people tend to sit in circles rather than stand and mingle; as Anne Knudsen has argued, this pattern emphasizes equality and belonging among the assembled party, even as it relegates anyone outside the circle to irrelevance (Knudsen 1996). Socially, too, being inside rather than outside carries enormous importance. To be an insider, a member of the group, is to know warmth and kindness and the untranslatable Danish experience of *hygge;* to be an outsider is to meet coldness and indifference, and even cruelty. Danish culture thus exerts a strong pressure on individuals to be part of the group, and it fosters a great fear of being set outside of it.

The paramount sin in this system is that of pride, of calling attention to oneself as different from and better than the rest of the group. Danes shrink from anything resembling self-importance, and they reserve some of their most scathing social sanctions for those guilty of it. This attitude comes through in the writings of the Danish author best known to the English-speaking world, Hans Christian Andersen; the suffering of characters like the Ugly Duckling embodies the venom directed at aspiring nonconformists in Danish society, a venom Andersen himself had known intimately (Andersen 1951). Danes express this attitude in terms of another author, the Danish-Norwegian novelist Aksel Sandemose, who created the archetypical town of Jante in *A Refugee Crosses His Tracks* (1936). Sandemose's "Law of Jante," the Jantelov, codifies the hatred of difference he saw in Danish rural culture; its ten commandments include "Thou shalt not believe that thou art something," "Thou shalt not believe that thou art as good as we," and "Thou shalt not believe that anyone is concerned with thee" (Sandemose 1936: 77–78).

The Jantelov is, of course, a literary creation, and it reflects Sandemose's own bitter childhood in rural Jutland. Yet most Danes agree that something like the Jantelov characterizes their culture, and they refer to it routinely when they describe attitudes toward self-importance. And not only self-importance—the Jantelov implies an attitude toward difference generally, toward any attempt to distinguish oneself as something apart from the group. To call attention to one's difference, to make oneself special, violates the principles of equality and sameness that Danes associate with the inside, and it invites a scathing response from the group. To be different is generally something undesirable in Danish culture, and to deliberately set oneself apart is tantamount to heresy.

As one might expect in a nation so concerned with group membership, Danes engage in an active discourse on the nature of group identity. As in many small national cultures (Jokinen 1994), Danish public discourse is very concerned with working out the details of what is Danish and contrasting them with what is not. Much of this discussion focuses on the activities of daily life—foodways, styles of speech and demeanor, language, housekeeping. Danish women's magazines, for example, invariably include articles on "real Danish" recipes and traditional Danish decorating; likewise, Danish flags and references to authentically Danish goods ornament everything from bakeries to hardware stores. Such references permeate popular culture in Denmark, even rock music. One of the most popular Danish bands during my first fieldwork was called Bamses Venner, and it played songs that sounded very much like American pop music of the day—thumping drums, loud guitars, and shouted lyrics in a driving rhythm. Its biggest hit at the time, however, had a chorus I could not have imagined coming from an American hard-rock band:[1]

Rød grød med fløde	Rhubarb pudding with cream
På en dejlig sommer dag	On a delightful summer day
Rød grød med fløde	Rhubarb pudding with cream
Er en dansk holiday!	Is a Danish holiday!

In addition to the customs of everyday life, conceptions of Danishness also draw heavily on national folklore, history, and religion. These three areas came together in the nineteenth century in the work of N. F. S. Grundtvig, a priest and poet whose integration of nationalism and spirituality profoundly influenced both the Danish church and Danish culture more generally (Balle-Petersen 1983; Begtrup 1936; Buckser 1995; Buckser 1996a; Nielsen 1955). Grundtvig argued that every people had a spiritual essence, a "folk spirit," which must be understood before true Christian faith could be achieved. That spirit could be found in the culture of the rural folk, as well as the mythology of the pre-Christian era. Gruntvig and his followers therefore advocated research into folklore and history as well as the political empowerment of the rural free farmers who embodied their image of true Danishness. Their efforts helped transform rural Denmark in the nineteenth century, and they continue to shape popular conceptions of history and Danish identity (Buckser 1996a; Buckser 1996b; Lindhardt 1959; Pontoppidan Thyssen 1957). The historical and folklore societies inspired by the Grundtvigians remain very popular in Denmark, for example, and the influence of Norse mythology can be found in everything from school curricula to the fine arts to baby names. To be Danish is in large part to share in this heritage, to see one's roots in the rural culture and ancient history of the Danish folk.

One other key element in Danish national identity involves religious affiliation. Denmark is a notoriously secularized country, one whose rates of religious participation have often been cited as among the lowest in the world (see Hamberg and Pettersson 1994; Salamonsen 1975). Few Danes avow an active belief in a well-defined God, and organized religion is widely regarded as rather archaic. At the same time, though, Denmark does have a state church, and the great majority of Danes belong to it. Known as the Danish Evangelical Lutheran Church, or the *Folkekirke*, the church claims about 90 percent of the Danish population as members. Though these members rarely attend a Sunday service, most of them do incorporate the church into their lives in one way or another; most are baptized and confirmed in the church, a majority are married in the church, and almost all eventually have funerals there. The church and its festivals, moreover, are deeply integrated into the national culture. Christmas is Denmark's largest annual event, with seasonal decorations and activities that consume the attention of the nation for over a month each winter. Other seasonal activities are also linked with church holidays, from major celebrations like Easter and Midsummer to minor festivals like Shrove Tuesday (*Fastelavn*) and St. Morten's Day. The images of priests and churches figure frequently in Danish speech and expressive culture as symbols of public authority and morality. For all its well-known rejection of religious belief, then, Denmark is in many ways a deeply Christian country, in which symbols of group identity are often inseparable from symbols of religious affiliation.

This style of national self-definition poses real problems for outsiders in Denmark, particularly members of immigrant minority groups. Such people often differ radically from Danes in the issues of custom, language, history, and religion that lie at the heart of Danish self-definition; their very presence violates the image of cultural uniformity so central to the Danish national self-image. Given the stigma attached to even minor deviations from group identity in Denmark, this sort of wholesale otherness can produce serious problems for those who embody it. Muslim immigrants from the Middle East, especially, have encountered suspicion, resentment, and sometimes open hostility from the larger society since they began to arrive in large numbers in the 1960s (Blum 1982; Blum 1986; Enoch 1994). Their efforts to maintain distinctive religious and cultural customs have met angry reactions from the larger populace; efforts to build a large mosque in Copenhagen collapsed under heavy criticism in the 1980s and 1990s. In recent years, as problems of unemployment have grown more severe in Denmark, Muslims have often been scapegoated by rightist political leaders, and calls for their expulsion have become familiar refrains of groups like the Danish People's Party and the Progress Party.

Such vilifications of foreigners have drawn justifiable outrage from immigrant advocacy groups, who point out their contrast with Denmark's widespread reputation for tolerance. Indeed, the treatment of foreigners has prompted charges of hypocrisy by many inside and outside of Denmark, who say that the anti-Muslim sentiment exposes the Danish reputation for tolerance as a self-serving lie. The issue is more complicated than that, however; the Danish reputation for tolerance has a very real grounding, one visible in the open attitude toward a variety of nonconformist social and personal practices in Denmark. The controversies represent the difficulty of incorporating certain kinds of difference in a society that sees itself as largely homogeneous. Where the distinction between inside and outside is so important, and where a particular set of symbols defines the nature of the inside, the presence of visible outsiders can be deeply discomfiting. Attitudes toward foreigners are shot through with ambivalence, reflecting the real tensions in a culture that prizes openness and tolerance even as it castigates difference.

PERCEPTIONS OF AND ATTITUDES TOWARD JEWS

My first encounter with Danish attitudes toward Jews came not in Copenhagen, but on Mors, the small island in northwest Jutland where my wife and I did our initial fieldwork. Mors has something of a reputation in Denmark, where it stands for rural xenophobia in the same way that Nashville stands for country music in the United States. Our friends in Copenhagen warned us darkly of the narrow-minded provincialism we would meet there, of how suspicious and backward our new neighbors would be. They were wrong, of course; Nykøbing Mors was perhaps the warmest and cheeriest place I have ever lived, and we were very sad to leave it. Still, our preconceptions lingered, and we were particularly nervous about the reaction they would have to my wife's religion. Here we were in rural Europe, the same rural Europe that had burned shtetls with such glee a century ago—what would they think of a Jew in the fold? How would that make them treat my wife, how would it make them treat our daughter? For the first six months, therefore, we kept that personal detail a secret.

One day, however, our elderly neighbor was visiting, a retired farmer from a small village in the south of the island. We had become quite close to Holger Vester, and when he asked something about my wife's curly hair, she very nervously mentioned that it ran in her family—they were Jews. Holger sat bolt upright.

"You're Jewish?"

"Yes," my wife replied, more nervous than before.

"Ha!" he said, with a broad grin. "I knew it! Ester said you were Italian, but I knew you were Jewish!" It emerged that my wife's ethnicity had been the subject of a lively debate among the people who knew us, and that the leading candidates were Italian and Jewish. Holger was thrilled and went home immediately to tell his wife; within a few days, everyone seemed to know about it. And now that it was out in the open, people talked about it a lot. Visiting neighbors asked about Jewish holidays and customs, including how Jewish families decorated their Christmas trees. Whenever a newspaper article appeared on a Jewish subject, we could count on our neighbors bringing it by; newspaper clippings also occasionally appeared in the mail, addressed to "The Americans in Grønnegade." In a small country town, where almost everyone belonged to the state church, Judaism was something unusual and interesting, and people liked to talk about it. They were uninformed, and they asked questions which in other contexts might have been offensive, but their questions were never hostile. Indeed, in the rest of our six months there, in the navel of Danish provincialism, we never found a trace of an insult or a social withdrawal relating to my wife's religion.

One wouldn't find the same reaction in rural Poland, or in the rural United States, or for that matter in the urban United States. Anti-Semitism is so much a fact of life in most of the world that its virtual absence in Denmark is difficult to accept. And yet that absence is real, a fact affirmed by almost all of my informants. There are stereotypes of Jews, of course, stereotypes that resemble those elsewhere in the West. Jews are associated with tightfistedness, cleverness, big noses, teetotaling, and clubbiness. But these associations lack the venom that accompanies them so many places in the world, differing little from the stereotypes that the French are snobby, or that South Jutlanders speak fast, or that people from Århus are a bit slow on the uptake. If anything, indeed, Jews tend to have a positive image in Denmark, where their association with the 1943 rescue makes them an object of national pride. The importance of the event to Danish national self-perception turned the Jews into a sort of mascot minority, a group whose continued existence attested to the worth of the larger population. In a self-consciously homogeneous society, moreover, which generally frowns on expressions of individual distinctiveness, Jewishness offers an acceptable sort of difference. Non-Jews with a Jewish grandparent or in-law refer proudly to the fact, as a sort of seasoning that distinguishes them from their blandly Danish neighbors. To be Jewish is to be different, certainly, but it is a difference that generally arouses curiosity and interest rather than fear or disdain.

These attitudes make Denmark an unusually safe place to be Jewish. When I asked my informants if they had ever encountered anti-Semitism in Denmark, most of them answered with a flat negative—they had never been attacked, discriminated against, or insulted by Danes because of their ethnicity. A few recalled moments when their Jewishness had been the basis of an insult, but the incidents were minor and unrepeated. One man, for example, had complained to a notoriously belligerent plumber about a faulty repair job. "I guess you come from a people that's used to having everything it wants," the man had retorted, a reference to the stereotype of Jews as rich. My informant stressed that the man had just wanted to insult him, and that Jewishness was simply a convenient target: "If I had had red hair, I think, he would have said something about that instead." Anti-Jewish incidents generally had this flavor, expressions of personal pique or spite rather than a basic antipathy to Jews. Organized anti-Semitism is virtually absent; while a handful of neo-Nazis publish occasional anti-Jewish pamphlets and web pages, they have no significant popular following and are generally dismissed as mentally disturbed. Even the rightist anti-immigrant parties, with their slogans of "Denmark for the Danes," have adopted no anti-Semitic rhetoric. While the fact of Jewish difference presents problems for living in Denmark, it does not raise the sorts of danger and exclusion that have scarred the Jewish experience through most of its history in Europe.

Beyond these general stereotypes and broad positive impressions, most Danes have very little contact with or knowledge of Jews or Judaism. Jews lack the presence in popular culture that makes them a well-known quantity in the United States; most Danes know Jews exist, but they know next to nothing in detail about them. I expected such unfamiliarity on Mors, but I was surprised to find it equally strong in Copenhagen, even among the educated. Early in my fieldwork, for example, I had lunch with a sociologist at the University of Copenhagen, a close friend of one of my Jewish informants. When I asked her what she knew about Jews, she found herself at a loss; the only details she could bring to mind were that there was something Jews weren't supposed to eat, and that they held their version of Christmas, which they called Pesach, in the spring. Words like *kosher, bar mitzvah, Rosh Hashanah,* and *schlep* had no meaning for her, nor could she identify any characteristic Jewish surnames. She thought that she could probably recognize a Jew on the street, but she could not say how, and she was surprised when I told her the names of a few prominent actors who were Jewish. Even Woody Allen came as a surprise; thinking back on his movies, she said she guessed he had mentioned being Jewish, but she had never thought of him as a Jewish actor. This general lack of awareness echoed in many of my interviews, as Jews told of having to explain the most basic elements of their faith and practice to non-Jewish friends and co-workers.

When they learn about Jewish practice, Danes generally make sincere efforts to accommodate it. None of my informants reported difficulties arising in the workplace as a result of religious observance; workers who could not attend meetings on Saturdays or who could not eat at non-kosher office parties usually found their co-workers sympathetic and anxious to make alternate arrangements. Their unfamiliarity with Jewish customs makes doing so difficult, and at times almost comical. Jews told me of arriving at dinner parties where the host had made sure to avoid serving pork, but had replaced it with shellfish or beef in a cream sauce. Similarly, one of my orthodox informants told me of difficulties in her high school, when she was unable to attend the Friday night dance that was to end the school year. Her friends were horrified and deeply concerned. When she told them she could not drive on a Friday evening, they offered to drive her. She couldn't ride in a car? They would walk to the dance with her. She couldn't buy drinks? They would buy them for her. She couldn't carry anything? They would hold her handbag. She laughed as she recounted their increasingly desperate efforts to include her in the event and her own inability to get across the basic impossibility of her participation. While Jews find a general sympathy with the strictures of their religious practice, they also find a profound ignorance that can make real accommodations impossible.

Jewish Views of Jewish Difference in Denmark

One of the questions I always asked during my interviews had to do with Jewish difference. Imagine yourself, I said, at a cocktail party, a large affair with perhaps fifty other guests. People are milling about, talking in little groups. In this room there is one other Jew, not someone you know personally. How long will it take before you know who it is? The answers I received varied, but not much. Some said they would recognize the person immediately; others said it would take thirty seconds; a few suggested it might take up to two minutes. None of them thought that finding the other Jew would be difficult. When I asked how they would be able to tell, my informants usually became rather vague. It was not one thing, they stressed, but a lot of little things that would give the Jew away, little features of appearance or mannerism that were hard to describe but impossible to miss. They usually laughed as they struggled to put into words what it was that made Jews different. In the end, many of them resorted to a statement on the order of "Maybe you have to be Jewish to see it."

Descriptions of Jewish difference among my informants generally had this character—there was almost uniform agreement that Jews differed significantly from other Danes but little articulated sense of what that difference was. Jewish difference was not so much believed in as felt, experienced as a recurrent feature of a lifetime of interactions with Jews and non-Jews. Many told me that they felt more comfortable with other Jews, that they felt an immediate emotional connection that they did not experience outside the group. I expected such statements from religious Jews; I was surprised, however, to find equally strong statements from many liberal Jews, including some who were intermarried and participated in very few Jewish activities. The depth of the difference came through in their terminology. While most of my informants were Danish-born, native Danish speakers, and deeply enmeshed in Danish society and culture, they tended to talk about "Danes" and "Jews" as two distinctive categories. To be Jewish in Denmark is to be feel palpably different, even if one cannot say precisely how or why.

Part of the distinction clearly involves physical appearance. Both Jews and Danes vary widely in physique; despite popular stereotypes, many Danes are short and dark, and many Jews are tall and fair. Most Jews, moreover, have intermarriage in their family history, and some have as much Danish as Jewish heritage. Even so, in a small country highly attuned to minor deviations from the norm, such features as curly hair or olive complexion are quickly noticed and remarked upon. One of my informants, a well-known actress, said that her "Jewish face" and small stature meant that she usually wound up in exotic roles; while I would have found it hard to identify her as Jewish, casting directors never did. She could play an Arab princess, or a gypsy fortune-teller, but she would never be cast as Juliet. Similarly, a number of my informants said that they were frequently taken for foreigners, particularly in tourist areas or at border crossings. Another woman, who had long black curly hair, recalled a very pleasant customs inspector complimenting her on her Danish. "He asked how long I had lived in the country, and I answered, 'thirty-seven years!'" While Jews are seldom teased or insulted about such physical differences, they are made aware of them.

Beyond appearance, Jews frequently referred to differences in affective style between themselves and other Danes. Jews, they said, are more direct, more immediate in their presentation of self. They come right out and tell you what they're thinking, rather than holding emotions in reserve like most Danes. Some of my informants characterized this pattern as confidence, while others (usually with a smile) described it as *chutzpah*. One told me that this was why Jews had a reputation for not drinking; unlike Jews, most Danes needed to have a few drinks in them before they could relax and express themselves. Another described her experience in attending a Jewish day school as a teenager: "All my life I had felt like there was a curtain between me and other people, that it was like talking on a long-distance telephone line, where there's a moment of transmission time between when you say something and when the other person hears it. Then when I went to the Jewish school, it was like being surrounded by people who could hear me, whom I could talk to without the filter. I said something, and they understood it and shot something back right away. It was wonderful!" Such affective differences can be important in a group whose members are normally on close, even intimate terms with non-Jews. In mixed marriages, for example, they can sometimes cause strains, not only between partners but among in-laws. Jewish parents sometimes find their children's spouses distant and uncommunicative, seemingly cold in a family setting usually associated with warmth and intimacy. Intermarried Jewish women, meanwhile, often told me that they thought they were a little hard on their husbands, that their brassy directness was more than what these phlegmatic Danes had bargained for. I never heard such a complaint from the husbands, but the issue clearly worried the women at least a little.[2]

One might interpret these statements as essentially made-up, as the projection by Jews of popular images and stereotypes onto themselves. Yet they are clearly more than that. If Jews see themselves as affectively different from other Danes, this is in part because many of them genuinely are. The differences are not immediately obvious; in my first visits to the community, I would have had difficulty identifying any distinctive Jewish affective patterns. Jews do not talk with a special accent, they do not gesticulate distinctively when talking, they have the same wry and cultivated style of self-presentation that impresses Americans about Danes generally. Over time, however, I noticed a number of subtler patterns that kept repeating themselves, and that made me understand, at least in part, why they were so sure that they could spot one another at a party.

Compared with other Danes, for example, Jews tend to touch more physically. In the synagogue courtyard after a service, one can see frequent exchanges of hugs and kisses; one might, if one were lucky, see a few handshakes in a similar setting outside a church. Children were more frequently kissed, hugged, and held in the Jewish households I visited; indeed, children were generally treated with a degree of indulgence unusual even in this remarkably child-friendly land. Jews also had a different way of speaking about themselves, one that lacked the diffidence I often encountered in interviews with Danes. Danes can often seem apologetic when called on to talk about themselves, reluctant to violate the Jantelov's injunction against self-importance. My Jewish informants seldom gave this impression, and indeed I more frequently met a positive enjoyment of self-presentation. People told me about the importance of work they had done, about their high test scores or incomes, about the quality of Jewish institutions and publications, with little trace of embarrassment—and sometimes, indeed, with a degree of exaggeration.

On at least one count, Jewish difference involved an intensification of a wider societal pattern, a sense of being more Danish than the Danes. This was their sense of closedness, the extent to which boundaries and inside/outside distinctions make up a central feature of community life. When I first began working among Copenhagen Jews, I was startled and rather disheartened by the vivid feeling of exclusion that emerged from my encounters with them. I had met such boundary issues before, of course, having worked with other Danish populations, and I thought I knew what to expect from them; after a period of exclusion and suspicion, I would break through to the inside of the group and experience the warmth and intimacy with which I had come to associate social life in

Denmark. That breakthrough never happened with the Jews, however. Individuals were often friendly, and in some cases they were extremely helpful in my research, but I always remained outside the group, a stranger making observations rather than a participant in the social round. I was very seldom invited into homes and almost never given the house tours I was familiar with from Jutland. At one point, while visiting the home of a community member, I apologized for being a bad guest, and received a response that crystallized my experience of fieldwork: "Don't worry, you're not a guest. You're a spy."[3]

I was not alone in this experience. As my fieldwork went on, I soon came to realize that many of my informants had precisely the same experience of the Jewish community, a sense of exclusion that left them feeling isolated and embittered. Polish immigrants felt shut out of the community by those who had been born in it; children of immigrants felt shut out by the old families; those who had not attended the Carolineskole felt shut out by those who had; those who had attended the Carolineskole often felt shut out by the key cliques. Only a handful of my informants described themselves as insiders to the Jewish world, people who felt included and at home there. Even they acknowledged the high boundaries around and within the community, and the difficulty of penetrating them. Jewish difference, therefore, involves for many a sort of double liminality, as they feel fully part neither of the Danish nor of the Jewish world.

This liminality is, for some Jews, a source of real distress; in a country as concerned with inclusion and belonging as Denmark, to be an outsider can be a difficult and painful experience. Almost all of the Jews I met, however, looked on Jewish identity as something valuable, an aspect of self they wished neither to deny nor to change. I saw none of the feelings of self-loathing sometimes attributed to Jewish populations in the West (Gilman 1986). On the contrary, even nonpracticing Jews spoke with pride of their Jewish background. Many described it as something particularly enviable in Denmark, where distinguishing oneself from the crowd is so difficult and dangerous; Jews, they said, were like a seasoning in a bland broth, and connections with them enriched a Danish family. As my actress informant put it: "I think that most Danes don't really know who they are, where they have come from. But if you are Jewish, you know just where you're from, and that gives you a kind of confidence that Danes don't have."

Jews did not denigrate either the mannerisms or the physical features that they identified as characteristically Jewish. While dark features or curly hair clearly marked them as different in Copenhagen, none of my informants referred to such qualities as unattractive, and I met few if any artificial blonds.[4] The only negative association with the Jewish physique I met was the caricature of the *yiddische mama*, a fat maternal figure many associate with Jewish middle age. One informant told me that young Jewish women were horrified at the prospect of turning into *yiddische mamas* one day; as a result, they worked very hard to keep slim. This observation may have explained something about my female informants, almost all of whom were strikingly thin. Even this caricature, however, had less to do with Jewishness than with age and maternity. The generally positive view of Jews in Denmark seems to find an echo in the Jewish self-image.

JEWISH DIFFERENCE AND PERSONAL EXPERIENCE

In some ways, Denmark is an easy place to be Jewish. The threats and discomforts of anti-Semitism are all but absent, and the imagery associated with Jewish identity is uplifting and positive. Jews can and do participate easily and publicly in almost all areas of Danish life. For all that, however, to be Jewish is still to be different, both in terms of heritage and of intuitive self-perception. And in Denmark, where belonging to the group carries such importance, being different is a difficult thing. Jews are, after all, also Danes, and they share Danish attitudes toward inside and outside. Like most Danes, most Jews long to be part of the group; yet to the extent that they affirm a Jewish identity, they define themselves as outsiders. Engaging with Jewish practice and the Jewish community, therefore, involves a sort of tradeoff, in which the importance of Jewish distinctiveness must always be balanced against the importance of Danish belonging. As they decide how to live a Jewish life, individuals must continually work to strike a balance between these two values and to manage the psychological strain that can often result.

This strain can be profound, as I learned in one of my first interviews in Copenhagen. I met Lars Salafsky, a middle-aged lawyer, at an outdoor cafe in Nørrebro, where he helped me manage the unfamiliar mechanics of a Danish coffee press. He handled the press expertly, as he handled everything; a slim, elegant man with a long face and dark hair, he exuded the quiet competence and easy self-confidence I had come to associate with upper-class Danes. That confidence faltered, however, as we began to discuss what it meant to him to be Jewish. He wanted, he told me, to be a Dane with a Jewish heritage, but that never seemed to happen. However Danish he tried to be, he always remained a Jew. "I am a Jew," he said, "because society makes me a Jew." He said it with a sort of sad bitterness, conveying a sense of both hurt and defiance at once. The experience of being Jewish, he said, involved perpetually being set outside, being always marked as just slightly different from the rest of the group. And in a society as sensitive to inclusion as Denmark, such subtle distinctions can hurt.

For Lars, as for a number of the Danish Jews I interviewed, the most difficult period of the year comes in December, as the nation enters the Christmas season. Christmas in Denmark, known as *Jul,* is more than a religious holiday; it is a general cultural celebration, a festive period that runs from the opening of December to the Feast of the Epiphany on January 6. Christmas decorations—cut paper ornaments, wooden figures of gnomes, images of fir trees and gifts—blanket homes and public places beginning on December 1. Jul references suffuse television programming and school curricula, while festive activities like the raucous "Christmas lunch" (*julefrokost*) occupy much of the attention of offices and factories. During the dark Danish winter, amid the snow and sleet and the painfully short days, Jul creates a bright spot that the culture embraces exuberantly. Such an embrace, however, creates problems for Jews. Most of them view Jul as an essentially Christian event; while many of the associated activities have little to do with church theology, the holiday's religious origins make it something incompatible with Jewish identity. Christmas in effect manifests the deep association of Danishness with the Christian tradition and thereby highlights an aspect of Jewish difference that can normally lie unobserved. It does so, moreover, at one of the high points of the year, a time associated more than any other with warmth, togetherness, and fun. At Christmas in Copenhagen, being Jewish can feel lonely indeed.

Parents, especially, can find Christmastime challenging. The imagery and activities of Jul appeal to children enormously, and many Jewish schoolchildren spend all day in schools and day care centers that celebrate the season. Naturally enough, many of them want decorations and trees at home, and parents have difficulty refusing. Many don't, especially those who have intermarried. One of my informants recalled a year when her daughter asked her to put up a Christmas tree. "But we're Jewish, dear," she replied. "Jews don't put up Christmas trees." "Of course they do!" retorted her daughter. "Jesper has one, and Camilla, and Aron, and . . ." She went on to list a dozen of her friends whose homes had Christmas trees. My informant held her ground, and no Christmas tree went up. Holding to Jewish tradition, however, comes at a cost of disappointed children and guilt-ridden parents. One woman told me that her WIZO group always spent its December meeting sitting around and ranting about their difficulties dealing with Christmas.

Lars has no children, but he still has his problems. One of the customs at his law office, for example, involves decorating the doors; a committee of the secretaries puts branches, paper ornaments, and other symbols of the season on every lawyer's office door. Knowing that Lars is Jewish, and not wanting to offend him, they never decorate his. "I know they're trying to be thoughtful," he told me, smiling bitterly, "but perhaps I would like to have something on my door once in a while." He also runs into trouble with the office Christmas party. A boisterous event at which alcohol flows freely, the party is always held on a Friday evening. For Lars, however, who observes the Sabbath, attendance at a Friday night party is an impossibility. Similar reminders of his difference occur throughout the season, and indeed throughout the year. People raise their eyebrows when they hear his last name; they leave him off lists for Christmas and Easter cards; they ask him questions about Israel as though he were an expert; they nod knowingly if he does well in an investment. None of these acts in itself amounts to very much. Taking them together, though, Lars hears a recurring message, a statement that however hard he tries to be Danish, he will never entirely succeed.

It would be wrong to overstate the impact of this kind of message—Lars is not a bitter person, and he does not live his life in torment over his Jewishness. At most times, like most Jews, he lives like an ordinary Dane, with plenty of friends and a good social network. He enjoys living in Denmark and plans to stay there. The tensions over his difference are not a dominant theme in his experience but a discordant note in an otherwise well-integrated Danish life. The note does not go away, however, and it is a sound to which Danish ears are particularly well attuned. Again and again in my interviews, my informants talked about their feelings of exclusion, and many conveyed a sense of resigned bitterness about them. The creation of self-identity among Danish Jews faces the same dilemma as does the rest of Jewish life in Denmark, the need to merge two systems that are incapable of being fully brought together. Such an effort causes strain, and here and there the strain shows through, whether as wistful sadness as the prospect of Christmas, or as tufts of crabgrass poking through the gravel paths of the cemetery.

7

The World of the Past

Jewish Identities
and the Rescue of 1943

*In every generation a person is obligated to regard himself as if he had come out of Egypt,
as it is said: "You shall tell your child on that day, it is because of this that the Lord did for
me when I left Egypt." The Holy One, blessed be He, redeemed not only our fathers from
Egypt, but He redeemed also us with them, as it is said: "It was us that He brought out
from there, so that He might bring us to give us the land that He swore to our fathers."*

These lines from the Haggadah embody a central theme of the
Passover seder, the ritual meal that commemorates the liberation of the Jews from slavery in Egypt. They call on Jews to experience their common past in a particularly intense way. The
distant past, they argue, is not something far removed, something of merely
historical or nostalgic interest; it is, or at least it should be, something as
basic to individual experience as the events of contemporary life. This
theme runs through much of Jewish scripture, which calls on Jews to incorporate the collective past into the individual self on a profoundly intimate level. To be Jewish, in this sense, involves a special relationship with
one's predecessors, whose trials and triumphs are not merely history or allegory but personal experience.

I doubt that many Danish Jews really experience the Exodus this way. Like
most Danes, they tend to look skeptically at rituals like the seder, and they
could hardly relive events that they do not believe actually happened. In other
ways, though, the past has a deep influence upon Jews in Denmark, as it does
on most Jews in the West. For to be Jewish in Europe is to live in the shadow
of a history, the history of Jewish relations with Christian majorities over the

course of two millennia. That history is not something distant or abstract but a central element of the way that Jews see themselves and their neighbors. Its brutality, for example, figures directly in the way that Jews understand their contemporary situations. If Moses' journey across the Red Sea is too mythical for Danish Jews to relive, the Inquisition and the pogroms and the Holocaust are not; they lie perpetually in the back of Jewish consciousness, a grim counterpoint to the acceptance and tolerance on the surface of contemporary culture. To be Jewish is not merely to keep kosher, to hold certain beliefs, to call oneself a Viking Jew or a Pole and play soccer with Hakoah. It is also to know, deep down, that history has not been kind to you, that the paradise of tolerance that surrounds you is not something that has historically been reliable, that sooner or later such idylls have always ended, and ended badly.

This chapter discusses the place of the past in formulating Jewish identity in Denmark. Jewish identity everywhere involves not only social and spatial components but also temporal ones. In their religion and their folklore, Jews refer to a historical narrative that stretches for over five thousand years, and they understand themselves and their culture in its light. The past is not a printed book, however, something engraved by events and now unchangeable. The past is an idea, a complex of stories and memories that achieves reality only in the minds of people in the present. Like all ideas, it is something alive, something that reformulates itself with each person who conceives it. It is as much a reflection of history, culture, and experience as anything else in the present (Buckser 1995; Handler and Linnekin 1984; Hobsbawm and Ranger 1984; Lowenthal 1985; Thomas 1992; Trevor-Roper 1984; Zerubavel 1994).

As we consider the role of the past among Copenhagen Jewry, therefore, we must ask not merely how the events of the past shape Jewish identities but also how Jewish identities shape the nature of the past. And as with most elements of the Copenhagen Jewish world, no single or simple answer to this question exists. There is no common "Jewish version" of the Jewish past, any more than there is a common understanding of the nature of Jewish identity or the meaning of Jewish ritual. Construction of the past is a profoundly individual process, one that incorporates the varying decisions people make about the relationship between Danishness and Jewishness, group and other, religion and practice, and so on. What unites the myriad individual visions is not the story that they tell but the common reservoir of Jewish events on which they draw.

In this chapter, we will look at how one such event is constructed: the rescue of the Danish Jews from German persecution in 1943. The rescue was an epochal event for Copenhagen Jews in a number of ways. It enshrined for them a particular place in postwar Danish culture; it changed their relationships with non-Jewish compatriots; it gave them a new distinctiveness and visibility in world Jewish culture. For individuals, the rescue was a transformative event, one that left many of them changed people afterward. It is a subject that has been explored in dozens of books, hundreds of scholarly papers, and thousands of press and popular accounts; each Jewish family, moreover, has its own story of what happened to it during the war. The rescue also has a powerful significance in the larger Danish culture, as a heroic moment that defined the role of Denmark in the postwar world. Of all episodes in the history of the Danish Jews, the rescue is probably the one that has been talked about and thought about most, and accordingly it makes a good place to look at different ways of thinking and telling the past.

This chapter discusses some of the ways that individual Danish Jews have made sense of the rescue and how their understandings are related to their experiences of Jewishness and Danishness in the contemporary world. It suggests what their differences mean, both for the nature of history and for the nature of the Copenhagen Jewish community.

Basic Background

We can begin with a brief reprise of the major features of the rescue.[1] The Germans had occupied Denmark in April 1940, capturing the country under the pretext of safeguarding it from British invasion. The occupation was a relatively mild one at first. Hitler saw Denmark, with its Nordic population and Germanic roots, as a natural ally; it was to be not a conquered country but a "model protectorate," a vision of things to come in the new Nazi world order. While Germany imposed a military occupation and a ruling plenipotentiary, it left Danish civil institutions largely intact. These institutions included the Danish Constitution and its electoral system. Germany abused its position, of course, raiding the Danish treasury and violating various civil liberties provisions. For most Danes, however, the first years of the occupation meant relatively little change in work or lifestyle.

From early on in the occupation, the status of the Jews became a symbol of this policy. The Danish Constitution unambiguously forbade discrimination on the basis of religion. The sorts of anti-Jewish policies that the Nazis had pursued elsewhere in Europe would abrogate the constitution and thus violate the legal system which the Germans had promised to respect. The Danish government consequently rebuffed all German demands that it register and restrict the activities of Jewish Danes. Even a request for a census of the city's Jews was essentially ignored, on the grounds that it singled out Jews for special treatment. The Jews became, in effect, a proxy for Danish self-government, making their protection a major concern for Danish officials.

The breadth of this concern was, in retrospect, quite striking. The government could have asserted its independence simply by protecting Jews who were Danish citizens; it had little legal obligation to illegal Jewish refugees, and indeed before the war it had treated them quite harshly. During the occupation, however, the government made no distinction between Danish and refugee Jews, refusing to take any measures against any of them. The courts likewise made a special point of protecting the rights of Jews. When a Nazi-backed hooligan lobbed a firebomb at the Great Synagogue in 1941, the court convicted him and handed down a startlingly severe sentence (Yahil 1969: 48–49). And indeed, concern for the Jews went well beyond the official channels of the Danish state. After the German invasion, the welfare of the Jews became a matter of widespread popular interest. Danish anti-Semitism had always been mild, but it practically disappeared after April 1940; Nazi propagandists bewailed the fact that even subtle anti-Semitic jibes had suddenly become taboo (e.g., Christensen 1943). Popular leaders echoed and stoked this sentiment, notably the Lutheran priest Hal Koch, who in speeches to Danish youth movements explicitly linked the fates of the Danish nation and the Danish Jews. While the government's protection of the Jews involved a point about constitutional authority, then, its policy also reflected a wider institutional and cultural interest in the welfare of Danish Jewry.

A number of legends have arisen in this connection about the role of the Danish monarch, Christian X.[2] Several suggest that the king prevented the introduction of the yellow star into Denmark. In one version, told that the Jews would have to wear yellow stars on their clothing, Christian replies, "If this is done, I and my family will be the first to wear them, as a token of the highest honor." In other versions, Christian does in fact wear the star on his daily rides through the city, prompting all Danes to follow his lead. None of these events actually occurred; the Germans never in fact tried to introduce the star to Denmark. The tales have endured, however, in part because such an action would not have been out of character for the king. Christian did become an important symbol of Danish independence during the occupation. His defiant horseback rides through Copenhagen were virtual parades, with flocks of cyclists and pedestrians accompanying him. And he did show a real concern for Jews. In 1942, Rabbi Marcus Melchior presented Christian with a copy of a book he had just published; Christian accepted it and sent a flowery letter of thanks, expressing his concern about a recent fire in the synagogue. The letter had a strong impact on the Jewish community, particularly in view of Christian's famous terseness in communicating with the Germans.[3] Throughout the occupation, Jews clearly felt that they had the sympathy of the crown, and indeed my informants today still expressed a great deal of affection for the royal family.

While the Jews were largely safe, the protectorate arrangement came under increasing strain in 1941 and 1942, and in the summer of 1943 it finally collapsed. Following a series of popular strikes and an escalation in resistance sabotage, the Germans imposed a state of emergency on August 29. The Danish army was disarmed, the government was dissolved, and the entire country was placed under German martial law. German forces rounded up potential subversives and interned them in a prison camp; these detainees included the MT's elderly rabbi, Dr. Max Friediger, as well as other prominent Jews. Shortly afterward, German plenipotentiary Werner Best began planning for the deportation of the larger Jewish population.[4] Special Gestapo troops were imported for the operation, a transport ship was brought to Copenhagen harbor, and plans were made for transshipment to the Theresienstadt camp in Czechoslovakia. The roundups were scheduled for the evening of October 1, the first night of the Rosh Hashanah holiday. Jews gathered for holiday dinners would make convenient targets; bands of Gestapo soldiers were to arrest them, bring them to the transport ship, and have them on their way south before the rest of Copenhagen awoke the next morning.

The operation failed spectacularly. Best's shipping attaché, Georg Duckwitz, leaked the plan to Danish government leaders several days in advance; they told resistance leaders and Jews, while Duckwitz arranged for the reception of refugees in neutral Sweden. Marcus Melchior announced the danger in the morning synagogue services on September 30, and by nightfall most of the Jews knew about the planned roundups. They turned for help to Christian neighbors, friends, and employers. In a remarkable contrast to the experience of most European Jews, they met overwhelming sympathy. Jews were hidden in basements, summer houses, and farms; large numbers found refuge in the main hospital, where doctors checked them into rooms under standard Danish names.[5] Resistance leaders scrambled to arrange passage aboard fishing boats from the north coast of Sjælland.

Within a few days, an underground network developed that managed to smuggle almost all of the Jews safely to Sweden. Some of them had family ties there, while others were boarded in refugee camps for the remainder of the war. Altogether, of the seven thousand or so Jews living in Denmark, only 481 fell into German hands.

Even the captured Jews fared better than most. Theresienstadt was a concentration camp, and its prisoners suffered terrible privations, but its purpose was not to kill them; in most cases, Jews stayed there for a limited time, then were sent on to one of the death camps.[6] The Danish government and Red Cross, however, petitioned the German authorities so energetically on behalf of the Danish internees that no further transshipment ever took place. The authorities even permitted a delegation from the Danish Red Cross to visit the camp, an obvious propaganda ploy that nonetheless temporarily eased conditions for the Danish Jews.[7] Finally, shortly before the end of the war, Count Folke Bernadotte persuaded the Germans to allow the evacuation of the Danish prisoners to Sweden. Fifty-one had died during their confinement; the rest were loaded into white buses, which transported them through the war-torn landscape to the north. They passed through Denmark on the way, and they were greeted with jubilant demonstrations along their route.[8] Altogether, fewer than one percent of the Jews in Denmark died as a result of the German roundups.

THE RESCUE: IMAGES AND CONTROVERSIES

In the years after World War II, the rescue of the Danish Jews grew into a widely told story, a dramatic narrative with a standard form that emphasized particular features of the events (Buckser 1998). In books, articles, and films, the rescue became an icon of Danish conduct during the occupation. Authors emphasized the Danes' rejection of Nazi anti-Semitism, their embrace of their Jewish countrymen, and the heroism and ingenuity of the rescuers. Denmark became enormously popular among world Jewry; foundations like Thanks to the Danes sprang up in America to commemorate the rescue, while in Israel the Danish people as a whole were listed in Yad Vashem's book of righteous gentiles. More generally, the rescue became a symbol of the ability of moral courage to overcome seemingly insurmountable obstacles. The legend of the king and the star spread worldwide; to this day, whenever I speak to a group of Americans about Danish Jewry, at least one person always tells me the legend as fact. In addition to its obvious significance for the Danish Jews, the rescue became a central element in the perception and self-presentation of the Danish nation as a whole.

A number of factors contributed to the story's prominence in the postwar world (see Buckser 1998). The rescue had real significance to the Danish state, helping position Denmark on the winning side after the end of the war. The cozy relationship between the Danish and German governments had led Churchill to describe Denmark as "Hitler's canary" in the first years of the occupation; when the war ended, Denmark's inclusion among the victims of Nazi aggression was by no means self-evident. The story of the rescue argued for a widespread opposition to Nazism that counterbalanced the government's often collaborationist policies. The story also had a powerful resonance for international Jews, as well as many others struggling to make sense of the newly revealed horrors of the Holocaust. The rescue offered a compelling parable for this discourse, a refutation of the argument that successful resistance to the Final Solution was impossible. If the Danes, occupied and virtually unarmed, could save their Jews by a collective effort, what excuse could others have for standing by? The political and moral implications of the rescue made it a popular and widely told story. As the story was retold, moreover, the politically and morally salient features of its characters became progressively more simplified and standardized. The Danes appeared as moral paragons, the Germans as evil monsters, and the Jews as helpless and largely passive victims. The complex and often messy history of the rescue settled into popular consciousness as a neat allegory.

The actual rescue, of course, involved a more confusing mix of actors and motives, which fit poorly into a simple allegorical scheme. While many Danes behaved with real heroism, many others shied away from participation in the rescue, and a few actively opposed it. Cruel as the Gestapo troops were, for example, they are often described as mild compared with the Danish Nazis who helped them with the roundups (cf. Rohde 1982: 80). Some Danes informed on Jews for a profit; even the heroes of the hour, the fishermen who transported the fleeing Jews, seldom offered their services for free. Some extorted fortunes from their customers, at least until the resistance established a standard fee scale. The Germans, meanwhile, played a much more ambiguous role than popular stories imagine. The Gestapo troops were ruthless enough in their search for Jews, but regular soldiers often looked the other way. Even the German commander, Werner Best, played a confusing role, simultaneously implementing the roundups and undermining them. He not only allowed Duckwitz to leak word of the impending operation, he also gave orders about sea patrols and the conduct of troops that enabled many Jews to slip through. Easy as it is to find heroes and villains in the story of the rescue, it is difficult to characterize the groups involved in any simple way.

These complexities raise some intriguing historical questions, and they have spawned an extensive historical literature. Why, for example, did Best play the role he did? He claimed after the war that he opposed the roundups, and that he was trying to minimize their effectiveness; many have suggested rather that he was playing a double game, trying to please Berlin while also distancing himself from culpability after the war. Other questions concern the logistics of the rescue, the astonishing concatenation of circumstances that made it possible for the effort to succeed so well. Some of the most interesting studies focus on the motivations for the rescue (see Buckser 2001). Why was it that the Danes, on the whole, acted so differently than did most other Europeans? Did they embody, as many have suggested, traditions of tolerance and democratic values deeply ingrained in Danish culture? If so, why were they so harsh to Jewish refugees before the war, and why did they fail to oppose the Germans on other issues of human rights during the occupation? The answers remain tantalizingly out of reach for historians to this day, making the study of October 1943 a particularly contentious historical field and leaving the old allegory as still the primary way through which most people hear about the Danish rescue.

That allegory, it should be stressed, has a lot of truth in it. The current scholarly enthusiasm for deconstruction has led to a number a revisionist studies of the rescue, each of which highlights the messy discrepancies that the popular story leaves out (e.g., Blüdnikow 1991). Authors have pointed to the outrageous fees often charged by the fishermen, or to the rejection of Jewish refugees before the occupation, and all but condemned the rescue story as a self-serving fiction. This is, I think, a serious mistake. The rescue was not the fairy tale that often appears in the popular press, but neither was it an ordinary event; even the most unromanticized tellings of the rescue convey a sense of moral drama that is impossible to ignore. Anyone who reads more than a few individual accounts of the rescue finds himself repeatedly stunned by individual details, fleeting words or actions that create palpable moments of grace. A very minor example: One woman told me about fleeing from a German patrol with her children and finding herself trapped in an alley closed off by a tall fence. Going back out to the street would mean being captured immediately, but within a few minutes she would be found there anyway. It was one of the darkest and most terrifying moments of her life. Some workmen on a building site above her noticed her predicament; without a word, they clambered down, used their tools to take down the fence, hustled her through, and hastily put it back up. They then hoisted her and the children into a rubbish bin, covered them with construction debris, and drove them safely away. All over Europe, at this same time, Christians were standing by as Jews were hunted down and killed, and in many cases they were eagerly helping themselves to the spoils. The matter-of-fact compassion, ingenuity, and willingness to act of these workmen—or of the Danes who handed housekeys on the street to Jews they had never met, or who crowded around Jews in trains to hide them from German soldiers—stand in such startling contrast with the rest of Europe that they demand a moral accounting. To study the rescue in its details, warts and all, is not to discredit it, but expose a deeper moral puzzle: that of how a clearly flawed people, operating from diverse and often impure motives, nonetheless collectively did something that was deeply and undeniably good.

Variations in Jewish Experiences of the Rescue

Harold Flender's influential *Rescue in Denmark* opens with an image of Rabbi Melchior, dressed in street clothes instead of clerical robes, standing before a puzzled throng in the synagogue the morning before Rosh Hashanah (Flender 1963: 17). "There will be no service this morning," he says. "Instead, I have very important news to tell you." He tells them about the roundups planned for that evening and urges them to spread the word. "By nightfall tonight," he says in closing, "we must all be in hiding." This compelling scene did in fact happen, and as Flender says, it played an important part in warning Jews about the German operation. What Flender does not mention, however, is the composition of the crowd that Melchior was addressing. The occasion was a regular morning service, not the holiday itself, and it attracted relatively few participants. Those participants did not represent a cross-section of the Jewish community. Many Jews in Copenhagen could not afford to spend part of a weekday morning in the synagogue, and many others didn't want to. Those who heard Melchior's warning came disproportionately from the well-off, well-connected portion of the congregation known as the Viking Jews. These people, and the friends and family members they told, received early notice of the roundups, giving them time to make the plans and arrangements necessary for a successful escape. Many others—especially the poor, the immigrants,

and the refugees—heard about the roundups much later and found themselves in a desperate scramble for safety rather than an orderly relocation.

Throughout the rescue and its aftermath, class and social networks played
important roles in shaping individual experiences. While almost all Jews were
saved, they were saved in different ways, with differing degrees of anxiety and
discomfort. These differences affected the ways that they spent their time in
exile and the world that awaited them on their return. No Jews that I interviewed really complained about the inequalities—how can one complain,
when millions of Jews in neighboring countries were murdered? They generally knew about them, though, and the differences affected the ways that
they understood and reconstructed the rescue in the years afterward.

Class influenced the rescue in a variety of ways. Early on, for example,
class strongly affected the order in which Jews learned about the approach-
ing danger. Duckwitz delivered his warning to Social Democratic leader
Hans Hedtoft, who spread it first to the MT president and subsequently to
other Jewish leaders; those Jews closest to the political and economic leadership, therefore, heard the news first (Flender 1963: 51–52). Other Jews
heard through professional networks and workplaces, giving an advantage
to those in steady or high-status employment. The news often came last to
those at the bottom of the community's economic ladder, immigrants or
refugees with limited contacts with non-Jews. Once they heard the news,
moreover, such people often had trouble acting on it. The well-off had
places to hide, non-Jewish relatives and friends with spare rooms or summer houses to put at their disposal. The poor, by contrast, often had no one
to call on, even if they had time to make a call. They were more likely to
find aid through the resistance, in overcrowded safehouses or hospital rooms
rather than the homes of friends. The high cost of crossing the Sound added
to the problem, with only well-off Jews able to afford the stiff fees charged
by the fishermen. While almost all of the Jews ultimately made it across, the
attendant anxiety and terror hit some much harder than others.

These distinctions continued during the Swedish exile. Most of the
Jews arrived in Sweden virtually penniless and had to struggle to find employment during more than a year of residence. Professional skills, business
connections, and Swedish family networks—all more common among the
Viking Jews—made that struggle significantly easier. Well-educated Jews
also qualified more easily for jobs in the agencies that resettled refugees
during the last years of the war. Again, none of my informants complained
about their treatment, but such opportunities created a real divide in
refugee experiences. A year spent in a rented house doing white-collar

work gave the rescue a different character than a year spent as a dishwasher in two rooms of a refugee shelter.

Nor was the return to Denmark in 1945 a uniform experience. Stories of the return make up some of the most moving moments in the popular account of the rescue, and for many Jews they marked one of the high points of their lives. After the terror of the evacuation and the rigors of exile, they returned to find their homes as they had left them, maintained and decorated by Christian neighbors. Some found their tables set with china, flowers, and candles, just as they had been on the Rosh Hashanah eve when the families had been forced to flee (e.g., Bamberger 1983). Many, however, especially the less affluent, returned to find their belongings vanished and new families living in their apartments. With their jobs long filled by others, their return to Denmark was a time of financial desperation. In the chaos following the German withdrawal, moreover, Denmark experienced a brief flareup of the anti-Semitism that had vanished so completely during the war, making matters all the harder for Jews on the economic margins.

Such distinctions, like most of the complexities of Jewish experience, appear very seldom in the rescue literature. While Jews stand at the center of the rescue story, they have figured surprisingly little in most retellings of the tale. Most of the action usually revolves around the rescuers—their motivations, their moral courage, and their ingenuity in devising rescue operations. The Jews appear primarily as cargo, relatively passive and faceless victims who benefit from the protagonists' valor. This tendency is perhaps natural, given the rhetorical uses to which the rescue has generally been put. But the Jews did have faces, and those faces had important effects on how they experienced the rescue. While all kinds of Jews were rescued, they were not all rescued the same way. Whether one was a Viking Jew or an immigrant, a religious Jew or a secular Jew had significant impact on how one escaped the roundups and on one's destiny afterward. The divisions that characterized the Jewish community in ordinary times, in other words, characterized it during crisis as well.

CONSTRUCTIONS OF THE RESCUE AMONG CONTEMPORARY DANISH JEWS

More than half a century has now passed since the rescue, and the world in which it took place has been transformed. Denmark and Germany are now close allies, common members of a European Union that shares arms as well as trading relationships. The story of the rescue has evolved as well, from a part of memory to a part of history, becoming something that most Danish Jews have heard about rather than experienced. When people tell about the rescue today, their accounts of it incorporate the years that have followed; they recall events and make sense of them in light of the complex personal histories to which they led. Their narratives, as a result, reflect the sorts of diversity and variation that have characterized Danish Jewry since the war.

This variation exists for those who took part in the rescue as much as for their children. When older Jews tell about the rescue, their stories reflect their often sharply differing experiences at the time. They also reflect their experiences in the years afterward, as they select details and characters that seem important in light of the people they have now become. This editing is not always conscious, and it is certainly not a form of misrepresentation. It is, rather, part of the ongoing process of reconstruction that inevitably accompanies our thinking about the past. Individual rescue stories also respond to portrayals of the events in the larger society. The narrative so common in popular discourse is one whose motives and shortcomings Jews know intimately, and in many cases they have constructed their own narratives in response to it. Most of the older Jews I interviewed had told their stories before, often to journalists or oral historians; when they did so, some of them had tried to correct popular misconceptions, while others had tried to cater to them. All of the Jews I interviewed, old and young, knew about recent histories by authors like Bent Blüdnikow, which have challenged the popular version of the events of the time. Their reactions ranged from enthusiastic agreement to bored dismissal; several pointed out with some impatience that the "revelations" contained in the books had been common knowledge among Jews for half a century.

The best way to get a sense of the variation in the ways Jews describe the rescue is to look at some examples. To that end, I now turn to four accounts of the rescue, told to me by Danish Jews in the summer of 1997.[9] I have chosen these particular accounts to suggest the range and diversity of the stories that are told about the rescue. They are not intended to be representative; they are chosen, rather, to give some sense of the intimate connection between personal experience and historical construction as Jews think about the events of October 1943.

Ingrid Nathanson

When I imagine the Danish Jews of the late nineteenth century, Ingrid Nathanson always comes to mind. A member of one of Copenhagen's oldest Jewish families, she lives in a gracious apartment in Sølvgade, across the street from Rosenborg Castle. Mahogany furniture and framed etchings lend an air of antique bourgeois elegance to the apartment; Ingrid's own courtly manner conveys the same sense, an old-fashioned formality remarkable amid the noisy bustle of downtown Copenhagen. A small slender woman, now quite frail, Ingrid grew up in this same area, in a large apartment above her father's china shop. She was born there in 1927. The world has changed a good deal since then, as has Ingrid's experience of the Jewish community. She describes the Judaism of her youth as a way of life, a rhythm that moved from Sabbath to Sabbath and informed all aspects of her existence. She came from a large family and had many Jewish friends living nearby. She attended a regular Danish school, which she finished after the war ended. Soon afterward, she married a young Jewish banker, but he died of a heart attack only four years into their marriage. Ingrid and her daughter moved back in with her parents, and she has remained in their house ever since, supporting herself through secretarial work. She is very close to her daughter, who has not married, and she is actively involved in the synagogue. Ingrid considers her faith an essential part of her life and identity; she dismisses "cultural Judaism" as a contradiction in terms, a denial of the religious basis on which a Jewish life must be lived. Since her retirement, she has spent much of her time reading scripture and studying commentaries on it.

When I visited Ingrid in her apartment, I asked her to tell me about the rescue in 1943. The story she told me embodies the image of the event idealized in Danish and Jewish history. Her father learned of the planned roundups during synagogue services on the morning of October 1. He quickly telephoned a Christian friend in the seaside village of Hornbæk, who offered to take in the family until the trouble was over. The family traveled there that day and avoided the roundups that evening in Copenhagen. Their visit was soon interrupted, however; two days after they arrived, they learned that their friend's home was to be commandeered as quarters for a German officer. Their friend quickly found them lodging in the countryside. They were received there warmly, despite their exoticness in the local context; Ingrid remembers the wide-eyed farm wife greeting them with the words, "Sådan ser I ud!"—"So that's what you look like!" The next day they took a boat to Sweden, and within a few days they were given a set of rooms in a converted Swedish elementary school. They remained there for the balance of the war. The family members all found jobs in the local area, Ingrid as a cook, her father as an accountant. When the war ended they returned to their old apartment, which the neighbors had cleaned and decorated with a profusion of flowers.

Ingrid described these events with the ghost of a smile on her face, as someone does who is telling a good story and knows it. For Ingrid, it *is* a good story. Not only is it exciting, it is also a demonstration of goodness itself. She dwells very little on hardship or evil; the presence of the Germans, for example, is implicit in the story, but she almost never mentions them directly. She mentions the excitement she felt when they arrived in Swedish waters, and she heard the voices of the Coast Guard calling, "Welcome to Sweden"; she does not, however, describe the difficulty and terror of the crossing beforehand, or the bureaucratic maze most refugees faced upon their entry. Her time in Sweden likewise appears as a gift rather than an obligation. She describes her job there as a learning opportunity, one that has come in handy throughout life, not as the grinding labor which it undoubtedly also was. And her return to Denmark is recalled in terms of flowers and hospitality, not of the wave of anti-Semitism that swept through Copenhagen shortly after the liberation.

She remembers these things, of course, and when I asked her she agreed that they happened. But while they may have existed in some objective past, they are not part of the way she represents the rescue now, either to herself or to others. Her rescue story is one of hospitality and heroism, of acceptance and courage by the Danes and Swedes. I mentioned the revisionist historical studies, and she bristled a bit; some people, she said, want to put out a certain story, and they are willing to twist the truth for effect. One can always find things to complain about, but who else did what the Danes did during the war? Where else in the world do the Jews have it so well? The story of the rescue is a story about the relationship between the Danes and the Jews, and the image it portrays is one that has been true to her experience of that relationship.

Hanne Goldstein

Hanne Goldstein is an ebullient widow in her seventies who lives in an apartment house in Hillerød. She was born in Poland and emigrated to Denmark with her family in the 1930s. After finishing school, she married a second cousin, Benjamin, who was a minor functionary in a large manufacturing firm in Copenhagen. Early on in their marriage, the couple lived above Benjamin's brother's tailor shop in inner Copenhagen, and Hanne split her time between minding their two children and working in the shop. Later they moved to suburban Hillerød, and as Benjamin advanced in the firm they became part of the town's comfortable middle class. Hanne regards herself primarily as a mother, and her apartment contains dozens of photographs of her children and grandchildren. She does not describe herself as a religious person; she does, however, belong to the synagogue, and she attends services on some of the holidays.

When I asked Hanne about the rescue, she resisted my attempts to characterize it in a general way. People had many different experiences, she said, and there was no single description that was true for everyone. She preferred to talk about it in terms of her own experience. Her own story is a dramatic one, and she has told it to a number of newspapers. She first learned of the roundups a few days in advance, from the wife of a German trade official. The woman was a regular customer at the tailor shop; she arrived one morning without any orders, chatted for a few minutes, and then urgently told Hanne that she had to leave town. Hanne now suspects that the woman's husband knew the Germans were going to lose the war and was trying to curry favor with the Danes in advance. Hanne was terrified. Since arriving in Denmark, she had spent her entire life in Copenhagen, and she had no idea where else she could go or how to get there. One of her children, moreover, had fallen down some stairs a few days earlier and now lay in the hospital with a fractured skull. She told her husband when he arrived home that evening. They decided that to appeal to the president of his firm for advice, and Benjamin went to see him the next day.

The president didn't think that the threat was real, but he valued Benjamin as an employee and agreed to do what he could. The next day, he summoned Benjamin back to his office. He told him that a house had been arranged for the family on an estate in northern Sjælland. Benjamin could work in the fields there, and the family would be safe from any trouble going on in Copenhagen. Benjamin was to take the train out to the estate, make the necessary arrangements, and then return for his family—by the time he returned, the false alarm might have blown over. Benjamin thanked him profusely and left, though he had no intention of following the plan. He and Hanne left the city together the next morning, taking both children despite the objections of the doctors.

Hanne remembers the train trip to the estate as one of the worst of her life. She had never been to the country before; their young daughter was dangerously ill; and their only guidance was a handwritten map drawn by the president the day before. To make matters worse, the train carried a large contingent of German soldiers, of whom Hanne was terrified. They finally reached the local station and made a difficult overland journey to the estate, arriving at the overseer's cottage in the dead of night. And as Hanne had feared, he had no idea who they were. The president's arrangements had yet to reach the people living there. Hanne therefore found herself in the middle of a strange, dark countryside, hungry and exhausted, her children ill and crying, with no possibility of returning home and only a baffled and suspicious bailiff from whom to beg shelter.

While the overseer was suspicious, however, he sympathized with the bedraggled and bewildered family, and he gave them a room for the night. Their treatment the next day, moreover, more than made up for the difficulties of the journey. When the overseer finally learned the details of their situation, he installed them in a beautiful cottage, immaculate and airy and far larger than their meager Copenhagen apartment. As they arrived there, they found a flock of farm wives cleaning the floors, airing out the rugs, and packing in fresh linens. Neighbors brought over baskets of food and clothing during the day, and they kept the home well supplied for the duration of the visit. The next day, the overseer put Benjamin to work on a fence-mending crew; Hanne took on the duties of a country housewife. The warmth and affection of their reception made a tremendous impression on Hanne, who describes the weeks she spent there as among the happiest of her life. She still visits the estate on her vacations and maintains close friendships there; her daughters think of it as a second home.

The idyll did not last, however. A man in a nearby village eventually heard about them and informed on them to the Gestapo for a fee. Soon afterward a detachment of Gestapo troops arrived at the local train station, asking for directions to the estate. The stationmaster sent them in the wrong direction, then alerted the estate owner, who told the Goldsteins that they were no longer safe. The estate owner arranged passage for them to Sweden through the resistance movement; to get there, however, they had first to return to Copenhagen by train and hide out for several days. On the way they encountered a problem. Since passengers into the city were inspected, the Goldsteins needed to travel in the baggage car. The conductor, however, had no desire to get involved with anything illegal, and he refused to put them there. He rejected first their pleas and then their offer of a bribe, until finally their resistance escort had to produce a pistol and level it at the man's head. He told the conductor that there would be no more discussion and that the conductor was to see that they arrived safely. If for any reason they didn't, the resistance would find out and kill him immediately. After this beginning, the trip went smoothly as the terrified conductor took pains to keep them safe and comfortable.

They reached Copenhagen safely and spent a harrowing three days packed into the hall closet of a friend's apartment. A Nazi sympathizer who lived next door caused them some anxiety, nosing around and asking pointed questions about all the groceries their friend was buying, but no Germans ever investigated. Finally they took a train to the north coast. At one point, while waiting on a platform, they found themselves next to a group of German soldiers, some of whom began playing with the Goldsteins' children. It was obvious to any observer that the Goldsteins were fleeing Jews, and Hanne had a few frightened minutes; it soon became clear, however, that these were regular German soldiers, not Gestapo, and that they had no interest whatever in apprehending Jews. The family reached the coast safely, was whisked efficiently into the hold of a boat, and had a quick crossing over to Sweden. They were housed in Sweden in a village for refugees, about which Hanne had no complaints. Things were more difficult when they returned to Copenhagen in 1945; someone else had moved into their apartment, and all of their furniture and belongings were gone. Benjamin still had his job, however, and in a short time they were back on their feet.

Hanne's story, like Ingrid's, is a rousing one, which in many ways embodies the adventure and the moral courage that runs through the standard story of the rescue. In other ways, however, her story is much more morally nuanced. The Danes she describes are not a uniform mass of moral paragons; they are individuals, motivated in some cases by moral principles, in some cases by practical concerns, and in some cases by craven self-interest. Nor are the Germans uniformly hostile. She speaks of the regular soldiers with some sympathy as young men far from home, more interested in playing with a pair of toddlers than in persecuting Jews. Her story is not a story of Danish nobility and German evil but of the mixture of good and bad, of hospitality and hostility, that characterizes Danes, Germans, and Jews alike. When I mentioned the recent studies by Blüdnikow and others, she smiled at me indulgently and chided me for my naiveté. Why was I surprised to discover that the Danes didn't behave nobly all the time? There are bad people everywhere, she said; "After all, the prisons are always full."

David Belski

David Belski is the fourth son of a Jewish weaver who came to Denmark from Poland in 1906. Like most weavers' families, David's was poor, and while his parents were religiously devout, they could seldom take the time away from work to attend the synagogue. David's brother followed their father into the trade after taking a regular education in the Danish public schools. David, however, demonstrated a gifted intellect at an early age, and after seven years in the Jewish school he attended the *realskole* in suburban Frederiksberg. At the same time, he learned to play the trumpet and became a competent jazz musician. He graduated in 1942, and after the war he worked briefly at an office. The jazz life had more appeal, though, and by the mid-1950s he had left his job and become a full-time nightclub musician. His musical talent, particularly his brilliance at improvisation, brought him quickly to the top of the field. During the 1960s, he formed his own band, which became quite well known within Scandinavia. He and his wife have a son and a daughter, both respected journalists now with children of their own. David does not consider himself a religious man, and his only practical tie to Judaism is that he still keeps kosher. Few of his close friends are Jewish, and it does not bother him that neither child married a Jew. At the same time, he identifies himself as unambiguously Jewish and maintains an active interest in Jewish subjects.

I asked David about the rescue during an interview in my office. David brushed aside questions about the nature or meaning of the rescue, but he was happy to tell me of his own experience. David was fourteen years old when the Germans invaded, a student in the Frederiksberg realskole. His classmates at the realskole showed him genuine concern during the occupation; the teacher admonished the class to stand together with its Jewish members, and even one classmate who left to join the SS Frikorps left David a note stressing that this was no reflection on him. When the roundups came, David's family was temporarily housed by some Christian acquaintances in the country, then transported by boat over the sound. While they were waiting for transport, David realized that they would need money, and he bicycled back to Copenhagen to pick up his wages from his employer. Shocked to see him, the employer handed him the money, told him he was crazy, and ordered him to go back into hiding at once. The family endured a harrowing passage from their safehouse to the coast, and then a stormy trip across the sound. The arrival in Sweden he remembers as one of the great moments of his life. The passengers had come onto the deck of the boat, and they were unnerved to be approached by a military ship as they neared the coast. When an officer shouted, "Welcome to Sweden" through a megaphone, however, the entire ship burst into cheers and spontaneously sang the Swedish national anthem. The family fared well in Sweden, and David found his life's work. Bored by a desk job, he fell in with some Swedes who needed a trumpeter for their jazz band. David loved the experience and soon organized his own quartet, which eked out a living playing for clubs and parties for both Danish refugees and Swedes. He also fell in love with the girl who would become his wife. By the time they returned to Denmark, the future shape of David's life had been determined.

David's story of the rescue is in some ways the classic story—he tells of supportive classmates, generous helpers, a tense passage, and a rousing welcome in Sweden. Yet it is also the story of an independent young man who finds life, love, and his future by striking out on his own. His narrative highlights this independence. It does not, moreover, define him exclusively or even primarily as a member of the Jewish world. Throughout the story, he interacts with non-Jews and even non-Danes, most of whom help him in one way or another. It is the Swedes in his first band who show him the path for his later life, for example, and when he forms his own band, neither its musicians nor its customers are exclusively Jewish. David's rescue story is the story of his own coming of age, his emergence into the world of his adulthood. And the world into which it brings him is defined not by religion or ethnicity, but by the love of performing.

Annelise Kaufmann

Annelise Kaufmann is twenty-six years old and has been a member of the Copenhagen Jewish Community all her life. She attended the Jewish school as a child, and as a teenager she participated actively in Jewish youth groups and sports clubs. After graduating from the state gymnasium, she spent a year living in Israel, along with other members of her youth group, and for a time she considered staying there permanently. She returned to Denmark, however, and enrolled in the fine arts program at the University of Copenhagen; since finishing, she has worked intermittently as a fashion photographer.

I met Annelise by appointment at a cafe near the university. We sat and talked over coffee about her experience of Judaism and the Jewish social world in Copenhagen. Her feelings about that experience were very ambivalent. Her parents had sent her to the Jewish school to give her a Jewish identity, and she had certainly gotten that; indeed, her involvement in it had quickly exceeded their own perfunctory affiliation. For all her participation, however, she never felt entirely accepted in what she regarded as the very insular world of the school and the Jewish clubs. She said that the atmosphere there was very inwardly focused, concerned almost exclusively with

Jews, Judaism, and the state of Israel. If one had a different viewpoint, one would find oneself painfully isolated, as she was. Her placement on a religious kibbutz in Israel reinforced this experience; she felt attacked by the intense pressure to follow a particular religious and political line. She still regards herself as "very Jewish," and she still attends services and keeps kosher, but she has distanced herself in recent years from Jewish social life. She has a non-Jewish fiancé, and most of her friends are non-Jewish students at the university.

About an hour into the conversation, I brought up the subject of the rescue and asked Annelise what she knew about it. She immediately sat forward in her chair and leaped on the topic, as though the proper subject of the interview had finally been reached. The rescue, she told me, is a myth; the story of the great heroic Danish effort to save the Jews has been exaggerated out of all proportion. There were only a few people involved, she said, and those who were got paid very well for what they did. The story is told not because it is true but because it covers up the really shameful record of Denmark under German rule. After the war, she said, the Danes did not want to be seen as collaborators with the Nazis; the myth of the rescue made them seem like heroes of the resistance. The Jewish community was happy to go along with the fiction, which made them a sort of national mascot of tolerance. And so the story had persisted on down to the present, nurtured by Danish national guilt and by the smug complacency of the well-positioned Jews. Annelise was glad that the true story was finally coming out in the press, and she referred me to a book recently published on the subject.

To some extent, Annelise's picture of the rescue does reflect recent trends in press coverage and historiography in Denmark. It also, however, reflects her own ambiguous position both within the Jewish community and in relation to the larger Danish society. Like a number of young Jews I interviewed, Annelise finds herself in an uncomfortably marginalized position with regard to the Jewish community. Throughout our conversation, she evinced a resentment of the congregation's inner circle, the religious Jews who attend every synagogue service and upbraid her for her lax observance. She called them hypocrites, people with "false religion," making a show of their devotion to impress the outside world. They had shunned her in school, they had attacked her in Israel, making her feel like an outsider because she didn't see religion as encompassing her whole life. It was these Jews, she said, who clung most tightly to the rescue story; they recounted it as a sort of primal myth and celebrated it over and over in memorial celebrations and anniversary volumes. In puncturing the myth of the rescue, Annelise was clearly trying to deflate their pretensions, to show their myth to be as hollow as their sham religiosity.

Yet at the same time, like so many Jews, Annelise does not feel entirely accepted in Danish society either. With her dark, wavy hair and dark eyes, Annelise doesn't look like a typical Dane, and she says that she can never forget that. In shops, salespeople immediately assume she is a tourist; they compliment her on her wonderful Danish and ask her where she learned it. Her university teachers always assumed that she was a foreign student, and several invited her to introduce herself to the class. Never can she be just like everybody else, participating fully and naturally in the only culture she knows. Her characterization of the rescue is in some ways a commentary on this situation as well. The great Danish story of tolerance and inclusion, the story that the Danes loved and accepted and even risked their lives for the Jews, is a for her self-serving fantasy, a deception. The truth, as she experiences it, is darker and crueler.

STORIES OF THE RESCUE AND NOTIONS OF JEWISH IDENTITY

As mentioned above, popular accounts of October 1943 tend to take a fairly predictable form, and to convey a standard set of moral and historical lessons. Jewish stories of the rescue, by contrast, come in a variety of different versions, and they imply sharply differing moral conclusions. For Ingrid Nathanson, the rescue is a story about Danes and Jews, one that reveals a relationship of tolerance and mutual support and that illustrates the depths of Danish generosity and the roots of Jewish gratitude. The characters in her narrative are homeless Jews and hospitable Scandinavians; the Germans appear only as part of the background, never as active or individual persecutors. The message Ingrid's story conveys is one of harmony and moral courage, and it has clear implications for the situation of Jews in contemporary Scandinavia. For David Belski, by contrast, the story of the rescue is less a social commentary than a *bildungsroman,* a tale of a young man coming of age in a tumultuous time. It lays the foundation for his own life on stage, in a world defined not by nation or faith but by an international fellowship of performers. Annelise Kaufmann's story of the rescue is not even really about the events themselves but about the way that Danes and Jews talk about them. For her, the rescue epitomizes the hypocrisy and self-satisfaction both of a nation and of a Jewish community that will not fully accept her. These stories defy abstraction into any typical "Jewish version" of the rescue story;

as in so many contexts for Danish Jews, their variety of viewpoint and interpretation are among their most distinctive features.

The past often plays an important role in self-definition for ethnic groups, offering a symbolic language through which to express ideas of common origin and substance (De Vos 1982). For Danish Jews, the rescue provides a particularly evocative setting for such expression; certainly at no other point in the twentieth century have Jewish identities in Denmark been so starkly and publicly at issue. Characterizations of the rescue are therefore, among other things, allegories for the relationship between Danes and Jews generally. Ingrid Nathanson's narrative presents a popular and idealized image of that relationship: she conceives the Jews as a self-contained religious community and the Danes as a culture of tolerance and moral courage. This view accords with the conduct of her own life, kept mainly within Jewish circles and focused around religious observance. Annelise Kaufmann, by contrast, has tried to live as both fully Jewish and fully Danish, and she has found the effort a frustrating one. Her appearance and her religious practice make her stand out among Danes, while her engagement with Danish identity keeps her out of what she sees as the Jewish inner circle. Her construction of the rescue story expresses her disillusionment with both communities, focusing primarily on the hypocrisy and self-congratulation she sees on both sides. As individual Jews come to their own resolutions of the conflicts between Jewish and Danish identities, they shape their understandings of the past to reflect the approaches they take.

CONCLUSION: JEWISH COMMUNITY AND THE MEANING OF THE PAST

As Danish Jews talk about the past, they talk about themselves and about their community. By imagining characters and events in particular ways, they express the conflicts, tensions, and identities that shape the worlds in which they live. They are not alone in this; over the past several decades, historians and anthropologists have explored the construction of the past in a variety of different cultures, with very similar results.[10] The past, they have found, is a surprisingly malleable thing, a symbolic world that is continually reshaped and reimagined to express ideas about the present. While there are limits to how far one can stretch the past, in practice histories can be understood much as Geertz understood rituals, as "stories we tell ourselves about ourselves."

The stories that the Danish Jews tell are distinguished, most of all, by their variety. If Jews were to tell themselves a story about themselves, it would necessarily be a story much like the one told in this book—a story of division, of opposition, of changing identities and shifting alliances. And since these changes and oppositions are implicated in individual identities, individual characterizations of this story would differ among individuals and over time. What would hold them together—what would define them as Jewish stories—would be not the narratives themselves but the circumstances that define them. They would draw, first of all, on the Jewish past, a set of events that, however constructed, involve people who called themselves Jews. In addition, they would reflect a characteristic set of issues and tensions that define contemporary Jewish life. In the stories recounted here, the narrators take a number of different approaches to presenting characters, to framing moral issues, to relating personal and society-wide issues. In one way or another, however, all of them address issues implicit in Jewish existence in Denmark: the nature of Jewish difference, the constitution of the Jewish community, and the conflict between Danish and Jewish dimensions of identity, among others. What Jewish constructions of history share is not a common group narrative but a common historical and cultural context, within which Jews create a profusion of individual understandings of the past. Like so much else in the Jewish world, history is not something that unifies Jews but something that expresses the shared foundation on which their disunity is based.

Conclusion

The Future of Danish Jewry and the Anthropological Study of Community

During my interviews with Jews in Denmark, I often asked about the future of the community. What would it look like in ten years? In fifty? Would it exist at all, and if so, in what form? The answers I received varied considerably, but a majority of them were strikingly bleak. Many Jews doubted that the Mosaiske Troessamfund could hold out for many more years, given its shrinking membership and its looming financial crises; other Jewish institutions, like Machsike Hadas and the kosher butchers, seemed destined for imminent collapse. My informants remarked on the high rates of intermarriage and the shrinking pool of religiously observant Jews. Given such trends, how could the community hope to survive? Some of the most pessimistic views came from older Jews, who sometimes found it hard to convey to me just how much the Jewish world had changed since their childhoods. Jewish life for them had once been something all-encompassing, an experience of the world in which everything moved in a rhythm from Sabbath to Sabbath, where Jewish observance and society had defined the contours of their existence. Now the Jewish community was hardly a community at all, just the remnants of a community, a collection of the people who maintained the old forms dimly and intermittently and who would soon cease to maintain them at all.

These views echo a persistent strain in contemporary Jewish social analysis, a belief that Diaspora Jewry is in the process of dying. Scholars from a broad range of viewpoints have pointed out the increasing fragility of Jewish communities in Europe and North America, both in cultural and demographic terms. The European Jewish population, already ravaged by the Holocaust, has fallen steadily since 1945 (Sacks 1994). The large Jewish populations of the United States, likewise, have eroded sharply over the past several decades, with little prospect of a rebound in sight (Della Pergola 1991). Beyond simple numbers, moreover, the salience of Jewish communities for their members has declined precipitously. The vibrant Yiddish subculture of shtetl Jewry has all but vanished, both in Eastern Europe and in the various communities where it thrived in the nineteenth and twentieth centuries. Jewish identity has less and less to do with individual social network, occupation, or worldview; Jewish organizations have become increasingly peripheral to their members' lives. To many observers, these trends herald the end of an age, and they leave contemporary Jews in a sort of limbo. In the words of Alain Finkielkraut,

> Judaism, for me, is no longer a kind of identity as much as a kind of transcendence. Not something that defines me but a culture that can't be embraced, a grace that I cannot claim as my own. In twenty years at the most, there will be no more than a handful of professional historians to tell us of the Jewish culture of Central Europe and the genocide that brought it to an end. We occupy that pivotal moment, that detestable moment, when our past enters into history. The last survivors of this civilization disappear, turning it into a sort of vague and bygone era, abandoned by the general populace and snatched up by the specialists. The Judaism into which I was born is increasingly acquiring the status of a historical object, marked by a sudden distance making it both a painful and desirable object of reflection. (1994: 89–90)

Other scholars, however, resist such a mournful view of the state of Western Jewry. In the decline of traditional Jewish institutions they see not a disaster but a rebirth, a release from the constricting forms of traditional Judaism and an opportunity to explore new Jewish identities. Some forms of Jewish practice, after all, have experienced a resurgence in recent decades; active Hasidic communities have blossomed in a number of Western cities, and several forms of Jewish mysticism have found new audiences. Emerging forms of Jewish involvement, from Reconstructionism to Jewish Renewal to gay and lesbian synagogues (Shokeid 1997), suggest that Judaism can flourish in new settings. To writers like Jonathan Webber, these patterns demand recognition as something other than the death throes of a tradition: "It would, I think, be ethnographically wrong to say that the Jewish tradition simply weakened in modern times; it is true that it has weakened, but that is only part of the truth. New solidarities have arisen, new militancies for new identities have taken shape, each of which may present only a fragmentary and parochial view of the Jewish totality, but collectively suggest that it is within a native awareness of the ethnographic complexities themselves that Jewish survival today is being enacted" (1994b: 85).

The Jewish world of Copenhagen shows little evidence of these "new militancies." Outside of Chabad House, no real Jewish alternatives have gained any traction over the past several decades. A recent attempt to create a new Reform synagogue has made little headway, and more radical notions like a gay synagogue are virtually unimaginable. Even so, the Jewish community has proven remarkably resilient over the past century, persisting into the new millennium despite frequent predictions of its demise. For all its purported disintegration, the Jewish community remains important to the interests and identity of thousands of people in Copenhagen who call themselves Jews. Reconciling this continuing vitality with the undeniable evidence of the community's disruption presents a real difficulty. How do we square the seemingly self-evident fact that the Copenhagen Jewish community has spent the last century on the road to collapse with the equally self-evident fact that it never seems to get there? And given these contradictory trends, how do we say anything sensible about its future?

One way to approach these questions begins with the nature of community. Whether it has declined or rebounded, the Jewish world has clearly changed from its form of two centuries ago; if we look at Jewish Copenhagen today and see a community, we clearly mean something different by the term than did Gedalia Levin or M. L. Nathanson. To understand the present and predict the future, then, we might sensibly begin by exploring what community is, to see whether there is a way of thinking about it that can encompass both the Jewry of the eighteenth century and that of the twenty-first. When we do so, I think that we will find an image of the Jewish future considerably more promising than the bleak outlook so prevalent in Copenhagen.

Models that predict the demise of Danish Jewry, like many academic models of Jewish assimilation, tend to imagine community in a rather artificial way. They imagine the traditional Jewish community as the face-to-face *gemeinschaft* of classical sociological theory, a group characterized by a unity of sentiments cultivated in a closed social environment. Such a community, whether Jewish or not, could never thrive in a late modern society; as geographic mobility and modern social systems disrupt traditional parochial identities, closed and unified social groups become all but impossible to maintain. If we equate Jewish community with this kind of a group, we inevitably consign it to oblivion. Throughout this book, however, we have suggested another way of conceptualizing the Jewish community—not as a particular group of people or beliefs but as a distinct body of symbols, mental resources through which individuals, in a number of diverging ways, make sense of themselves and each other. Jews constitute a community not because they think about Judaism the same way or because they share a particular blood or class relationship, but because they all construct their notions of individual identity on a common symbolic base. This base, formulated over the course of two millennia of Diaspora existence, has shown a remarkable ability to retain relevance and vitality despite the dizzying social transformations of the last century. There are good reasons to think that it will keep doing so—and that the Jewish community, for all its changes, will maintain its existence and its significance through the further transformations that surely lie ahead.

As we conclude our discussion of the Danish Jews in this chapter, I would like to say a bit more about this conception of community, its implications for the Jewish future, and its implications for the study of modern cultural processes beyond the Jewish world.

THE COMMUNITY AS SYMBOL IN CONTEMPORARY COPENHAGEN

The concept of community begins with commonality, with the notion of something shared among a group of people. The Copenhagen Jews clearly have such a sense about themselves; although they recognize the profound differences that separate groups of Jews from one another, they nonetheless see something common to the whole that distinguishes them from the larger non-Jewish society. The acknowledged difficulty in defining that shared essence makes it no less real to individual Jews. If anything, its elusiveness seems to testify to its profundity, making it appear that the bond among Jews transcends ordinary social relationships. Jewishness feels like something bred in the bone, something that cannot be reduced to an epiphenomenon of transient qualities like class or occupation. It has the character of what Clifford Geertz refers to as "primordial attachments"(Geertz 1973: 259).

That attachment is, moreover, essentially inexpungeable; whatever it is that makes a man a Jew, he cannot wholly escape it, even if he renounces the religion and binds himself into the non-Jewish social world. He will then be a lapsed Jew, or an apostate Jew, or perhaps a self-hating Jew, but he will never cease to be a Jew at his core. The shared essence that defines Jewish identity, in the eyes of those who avow it, has a force that individual desires can neither create nor evade.

Looking at Copenhagen Jews from the outside, however, it is surprisingly difficult to find anything in particular that they do all share, apart from the fact of being Jews. It is especially difficulty to find any common property that could exert the kind of binding force that Jewishness seems to have for its holders. On a formal level, for example, Jews do share a common religion, but they apprehend that religion in a striking variety of ways; as we saw in chapter 2, the Jews include both devout believers and outspoken atheists, ritual traditionalists and ardent reformers, as well as a substantial number of people who neither know nor care much about religious worship whatever its form. Their unity as a religion is not one of common belief or common sentiment, or even common behavior. Blood ties, likewise, form only a tenuous link among Jews in Denmark. While religious ideology asserts that all Jews are genetically related, the diversity of the Copenhagen community means that these links are often extremely distant, if not fictitious. The common ancestor that links a Yemeni immigrant and a Viking Jew may lie centuries or millennia in the past, hardly a commonality likely to inspire solidarity. Indeed, given the high rates of intermarriage in Copenhagen over the past century, many Jews have more and closer relatives among Danish Christians than among Israeli Jews. Similar points could be made for language, residence, appearance, or any of the other bonds we usually associate with primordial attachments. A search for some basic constitutional essence that characterizes all Jews in Copenhagen ultimately turns up nothing.

This situation derives in part from the individual agency implicit in constructing Jewish identity. In modern societies like Denmark, religious identities and behaviors are not imposed by any external authority. Individuals decide their styles of worship and affiliation by themselves, tailoring these styles to their own particular tastes and social networks. As modern social processes disrupt traditional residential, occupational, and other parochial social groups, these tastes and networks become increasingly eclectic; people construct their identities in relation to a hodgepodge of unconnected relationships and social networks, so that even people in the same family or household often have sharply different social worlds. Under such conditions, maintaining any uniformities of belief, practice, or sentiment among a group as diverse as Danish Jewry would be impossible. In a society in which every person is free to construe the meaning of Jewishness, no binding common identity can last. This situation is not, moreover, unique to Jews. Many authors have suggested that community in general withers under the onslaught of modernization. As the autonomy of individuals increases, the commonalities that hold any group together become more and more difficult to maintain. Over time, say many scholars, modernization involves the supplanting of cohesive communities and integrated identities by amorphous collectives and fragmented selves (see Appadurai 1990; Giddens 1991).

As Anthony Cohen has observed, however, modern societies are not the only places where people have different interpretations of their groups (Cohen 1985: 11–38). In every culture, in every community, behaviors and ideas about society vary from person to person; nowhere, not even in the remote island societies associated with classic anthropological community studies, does everyone in a culture attribute the same meanings to their common symbols. What they share are not thoughts or sentiments but the symbols themselves, a common set of signs through which their disparate ideas can come together. Indeed, says Cohen, part of the value of symbols in any culture is precisely their ability to unite people despite such differences. Two Americans saluting their flag may see very different things in the piece of cloth, one associating it with a tradition of military dominance and the other with a tradition of countercultural dissent. Such differences in understanding do not make their common nationality impossible; on the contrary, the ability of a single symbol to connote both of these meanings allows them to bridge their differences, to assert a common identity despite their conflicting views. Cohen argues that the classic image of a community in sociology and anthropology—a unified *gemeinschaft,* a localized group of related people who differ little in their understanding of the world—is essentially a fiction, a theoretical ideal that characterizes no real society. What does exist, what constitutes a real community, on Dobu or in Copenhagen, is a group of people who share a common set of symbols that they agree is representative of their collective self. At its base, then, a community is neither a matter of common essence nor of common understandings, but of commonly acknowledged symbols.

And at this level, the Danish Jews certainly share something important. For all their differences in religion, heredity, and worldview, they all have reference to a common symbolic base as they formulate their various identities. The Jewish tradition, originating in antiquity and developed over two thousand years of Diaspora, provides a reservoir of ideas, icons, ritual forms, linguistic formulas, and cultural patterns through which individual Jews can make sense of themselves and each other. The nature of Jewish history has made this symbolism a particularly rich resource; it touches not only on religious forms but also on questions of social order, of administrative procedure, of diet and dress and language and innumerable other areas of human life. In Jewish religious scholarship, it also provides an extraordinary analytical framework, a tradition of symbolic explication and debate that offers an endless fund of alternative interpretations. Individual Jews make very different things out of this tradition. Some venerate it, some excoriate it; some focus on its spiritual elements, some on its ritual practices, some on its social customs. Individuals pick and choose from this body of symbols as they work out their own definitions of what Jewishness is and what the Jewish community should be. Indeed, much of the day-to-day drama of the Mosaiske Troessamfund involves contending positions on which elements to emphasize. The process produces radical differences in belief and behavior, with Jews ranging from Danish-identified secularists to the stringent separatists of Chabad. Each of these interpretations, however, departs from the same common store of knowledge, the same shared body of symbols, and it is this that makes them all expressions of a shared Jewish community.

This notion of community allows us not only to encompass the diversity of Jews in contemporary Denmark but also to conceptualize the continuity of the community over the past three centuries. The Mosaiske Troessamfund dates its inception to 1684, and writers commonly describe the community as having existed uninterrupted since then. Yet few of the "primordial" elements generally associated with community have endured in any consistent form over that period. The structure of the MT itself has undergone several radical reformulations, including a fundamental reconstitution in 1814. Continuities of descent have been slim at best, as major population influxes have periodically reshaped the community; while one can find descendents of families from 1684 among the Danish Jews, it is far easier to find descendents of the Eastern European immigrants from 1900 to 1917, or of Polish immigrants from 1970. The languages used by Jews in 1684 have disappeared almost completely from Denmark, and their style of religious observance has little in common with the practice of most Danish Jews today. If we think of community in these terms, as a

matter of common blood or language or custom, then we have little reason to say that the Jewish community of 1684 still exists. What has endured, rather, has been the background against which all of these things are
defined, the body of symbols and knowledge which both Meyer Goldschmidt and Jacques Blum would immediately acknowledge as the Jewish
tradition.

This is not to say that the tradition itself has remained unaltered. Like
all human creations, the symbolic foundation of the Jewish community
changes over time, as the forces of history and human creativity infuse it
with new images and meanings. Some of these come from particular
events. The establishment of the state of Israel recast the meaning of Zion
within the Jewish religious framework, for example, while the nation's turbulent history over the ensuing decades has created a host of new understandings of Jewish identity and peoplehood. The Holocaust likewise
created new symbols of the Jewish experience even as it destroyed many
Jewish cultural resources. In addition to such distinct events, changes in the
larger world have altered the context within which Jewish life has taken
place, and in doing so they have altered the ways in which the symbolism
of the Jewish tradition has been used and understood. The rise of the modern state, for example, led to the transformation of Jews into citizens across
Western Europe, a change that gave Jewish social and cultural patterns new
and very different meanings. As both Jews and the world around them
change and evolve, the Jewish tradition develops as well.

Looked at this way, the question of the Jewish community's future takes
a somewhat different form. It is not a question of whether enough "real
Jews" will remain in Copenhagen to keep the synagogue open, or of
whether a pure Jewish tradition can resist the onslaughts of contemporary
culture. It is a question, rather, of whether the Jewish tradition will continue to be central to the way that significant numbers of people in Denmark understand themselves and their world. More importantly, it is a
question of how they will interpret that tradition and what they will contribute to it. It seems quite clear to me that in the short term, at least, the
Jewish community is in no danger of disappearing. Whatever the problems
of its institutions, Jewishness clearly has a powerful role in the self-definition of thousands of Danes who call themselves Jews, and the appeal of
ethnic identification in contemporary Western society is growing rather
than shrinking. What is less clear is the form that its survival will take. What
will people in fifty years, or a hundred, mean by calling themselves Jews?
How will Jewish identity relate to the other identities that they will nec-

essarily combine with it, and how will the Jewish community relate to the larger culture of Denmark? Rather than asking if the Jewish community will survive—I think the answer is clearly yes—I would prefer to ask what the Jewish community will become.

Much of the answer must await further developments. The major events that have shaped the Jewish experience over the past century have not followed any clearly visible path; no observer in 1900, for example, could confidently have predicted the destruction of European Jewry and the establishment of the Israeli state. The next century will bring its own history, and speculations about its effects on the Jews are pointless. We can suggest something about the future, however, by looking at the larger social context within which the Danish Jewish community exists.

THE JEWISH COMMUNITY IN THE DANISH CONTEXT

No community, Jewish or otherwise, exists in a vacuum. A surrounding world always exists, a context that structures the conditions under which common symbols are encountered and understood. Throughout the Diaspora period, Jewish communities have tended to experience this surrounding world in a particularly intense way. Most Jews have lived in small groups amid much larger populations, for whom they have often functioned as economic and cultural middlemen. This proximity has given the larger society a powerful role in defining the ways in which Jewish identities are apprehended and experienced. Societies have defined differently the settings in which Jewish identity is relevant to social action; they have imposed different political and economic consequences on Jewish affiliation; and they have offered different kinds of alternatives to compete with the symbolic world of the Jewish community. Jewish history, then, is in part the study of the Jewish context, the setting in which Jewish community is defined and enacted. And as we look to the Jewish future in Denmark, we may do well to briefly discuss how that context has changed over the community's lifetime.

As discussed in chapter 1, the Jewish story in Copenhagen begins in the seventeenth century, when Jewish merchants from Germany and Holland began settling in the growing metropolis. As the community developed, it did so in a context which largely excluded Jews from most of the Danish world. Except for business contacts and a very few social encounters, Jews lived their lives among Jews, speaking a distinctive set of languages and ordering their lives around traditional Jewish law. The Danish guilds, professions, and educational systems excluded Jews; in a nation with a strong state

church, moreover, Jews were not only foreigners but infidels, making marriage with them impossible and other relationships unlikely. At the same
time, however, the outside culture left the internal workings of Jewish society largely alone. Except for some restrictions on Jewish religious activities, the state meddled very little in Jewish life. For all their proximity to
the larger society, therefore, the early Danish Jews constituted an unusually
circumscribed group, with very limited and clearly identified contacts with
the non-Jewish world.

In this setting, the symbolic resources of the Jewish community functioned as the basic framework of the Jewish social world. Daily life took its
shape and rhythm from the Jewish tradition; social interactions operated
according to Jewish law, and spiritual life revolved around Jewish religious
symbolism. The disputes and divisions within the community revolved
around how this symbolic framework was to be interpreted and employed.
Which *minhag* should be used in services, the Polish or the Bremenite?
How should Jews behave at funerals, and which direction should the headstones in the cemetery face? Such disputes were protracted and bitter, even
when the Jewish community consisted only of a few families who knew
each other well. Whatever their differences, however, Jews expressed their
oppositions in terms of a common symbolic system, one that encompassed
virtually all of their experience. Going outside of that world—looking to
Danish law to resolve disputes, or seeking spiritual comfort in Christian religion—was virtually unthinkable for most Jews. The very few who did
so—those who converted to Christianity—were regarded both by themselves and by their Jewish families as having left the Jewish community.

The community's environment changed radically during the Reform
period of the early nineteenth century, especially after the Decree of
March 29, 1814. When King Frederik conferred citizenship on the Jews,
he ended the autonomy with which Jews had managed their internal affairs. The practical force of Jewish law disappeared as Jews became subject
to the legal and administrative mechanisms of the Danish state. At the same
time, a variety of programs tried deliberately to end the Jews' cultural isolation, ranging from Jewish schools that taught Danish to the forcible
opening of the guilds. The religious infrastructure of the Jews was re-
ordered along the lines of Danish religious organizations; the synagogue
took on the organizational form of a church, with a royally appointed chief
rabbi and new ritual requirements that echoed Protestant practice. Taken
together, the reforms transformed the meaning of the body of symbolic resources that constituted the Jews as a community. The Jewish tradition lost

its overarching role in defining the world of the Danish Jews; it became instead something partial, an element among others—government bureaucracies, Danish popular culture, and an increasingly diverse set of occupations and social networks—through which Jews made sense of themselves and the world.

The transformation did not take place overnight, of course, and the incorporation of Danish Jews into the larger Danish society was far from total. Religious affiliation still counted for a lot in Denmark after 1814; as a result, even though it had lost much of its administrative significance, Jewish religiosity remained an important factor that separated Jews from their Christian neighbors. Anti-Semitism remained important as well. While Denmark had never cultivated the vicious anti-Semitism endemic to much of Europe, strong prejudices and stereotypes were attached to Jews that made their full incorporation into Danish society very difficult. Indeed, the only large-scale violence against Jews in Danish history occurred in 1819, five years after the signing of the decree. Perhaps most importantly, the legal reforms had little immediate effect on the social and kinship networks within which most Jews lived their daily lives. While Jews became legal Danish citizens, their social interactions remained primarily Jewish, they remained concentrated in a small set of occupations, and they tended to marry other Jews. The Jewish world, while no longer circumscribed by the Danish legal system, remained a relatively restricted one, whose informal social and cultural boundaries held some of the same force that the old formal ones had.

The persistence of these boundaries meant that the nineteenth century saw a gradual shift in the character of the Jewish community, not a sudden transformation. Jews migrated gradually into non-Jewish social and occupational networks, and religious practice was only slowly decoupled from the daily experience of individual Jews. Over time, however, these processes profoundly changed the role of Jewish tradition in the lives of its members. Many of them drifted out of the community altogether as Judaism became increasingly peripheral to their lives, and as the hindrances it imposed came to outweigh the value of its symbolic meaning. Those who remained became divided in their understanding of its significance. Some rejected the Danicization of Jewish religiosity, flocking to traditionalist synagogues and emphasizing the importance of ritual practice and ingroup marriage; they tried, in effect, to use intense religious commitment to shore up the community boundaries which the legal structure no longer maintained. Other Jews, the majority, adopted the more limited view of Jewishness which their incorporation into Danish society encouraged. They adapted ritual practice to accord with a Danish lifestyle, they embraced Danish cultural and aesthetic forms, and they gradually relaxed prohibitions against intermarriage. This division between orthodox isolationists and liberal integrationists carried over into the twentieth century, and it remains an important current in the community at the dawn of the twenty-first.

The Jewish community underwent another sea change in the first part of the twentieth century, as Jewish refugees flooded into the country from Eastern Europe. The change occurred mainly in the content of the Jewish community, not its context; in effect, the immigration brought a wealth of new meanings into the Jewish world, new Jewish symbols and new interpretations of existing ones. The new immigrants came from a variety of Eastern European regions, each with different customs, linguistic patterns, and styles of ritual observance. Their class position and occupational backgrounds differed sharply from those of the older Danish Jews, and they often espoused political views unthinkable among the community's bourgeois leadership. As a result, the Jewish world became something both richer and much more fragmented as strong divisions of culture, class, and politics emerged among Jews for the first time since the 1700s. The framework within which the Jewish community existed, by contrast, changed relatively little during the time. Jews did become more visible, and images of bearded Eastern Europeans joined those of bloated capitalists in anti-Semitic stereotypes. The immigrants also changed Jewish relationships with state authorities, as the MT found itself in the role of a relief agency and advocate for the poor. On the whole, however, the context within which Jews related to their common symbolic tradition remained much the same as it had since emancipation.

A more profound shift came during and shortly after World War II. During the German occupation, the Jews attained an unprecedented visibility as symbols of Danish independence. Danes not only knew about Jews, they supported them, and the mild anti-Semitism of earlier years virtually disappeared. The events of the rescue established this symbolism more deeply, as Jews came to represent Danish resistance to the Nazis and as their rescue made Denmark an international symbol of moral resolve. This symbolism played a significant role in Denmark's postwar political position, and it became an important element in the imagery of Danish national self-consciousness. As a result, Jews found themselves after the war in an almost unprecedented situation. They were, for the first time since their arrival in Denmark, positively valued and enthusiastically embraced by the larger society. The barriers to full social integration largely fell away, and the Jews' legal equality was finally matched by a genuine cultural acceptance. In effect, the rescue completed what the Decree of 1814 had begun—the dismantling of the legal and cultural framework that had defined the Jews as a separate social group in Denmark.

The second half of the twentieth century saw a number of developments in the world of Jewish Denmark. The establishment of the state of Israel, for example, redefined the international context of Jewish life, and the new nation's changing fortunes had important effects on Jews' self-perception and public image. The Polish immigrants of the early 1970s produced important changes in the community, as have more recent immigrants from Israel and America. We have discussed the impact of some of these developments in the chapters above. One change that we have said less about, and which bears mentioning here, is a more general change in the meaning of ethnicity in Denmark, one that stems less from particular events or policies than from broad patterns of modern social change. This trend, which has characterized many societies in the contemporary West, may help account for the continued vitality of the Jewish community despite the erosion of its social boundaries.

Herbert Gans has argued that ethnicity undergoes a fundamental change in late modern societies, from traditional "social ethnicity" to something he calls "symbolic ethnicity" (Gans 1979; Gans 1994). Social ethnicity occurs in ethnic communities that constitute working social groups. Localized by neighborhood or occupation or religion, such groups have institutions, cultural forms, and social networks that largely encompass their members' lives. The Danish Jews of 1750 would have constituted such a community; so too might a Hasidic enclave in New York's Crown Heights today, or an Italian neighborhood in Boston's North End during the 1930s. When most Americans think about ethnic groups, they imagine settings like these. Social scientists, too, tend to define ethnicity using these communities as their paradigm and to measure ethnicity's strength by the extent to which such communities persist (e.g., Dreidger 1980; Goldscheider and Zuckerman 1984; Kivisto and Nefzger 1993). In the western world, however, their long-term endurance is all but impossible. The social processes associated with modernity—bureaucratization, individualization, the breakup of local and other parochial identities—make bounded communities of any sort difficult to maintain. The omnipresence of mass media and national culture, moreover, tends to undercut in-group languages and subcultures. As a result, social ethnicity tends to erode over time in almost any modern social group. In the United States, the various social institutions associated with what sociologists call "white ethnics" have declined steadily over the past half century (Alba 1985; Alba 1990).

This trend does not mean that ethnicity disappears. On the contrary, Gans argues (1979), the disruption of traditional groups makes something as apparently primordial as ethnicity all the more desirable (see also Alba 1990; Buckser 1999a; Buckser 2000; Jenkins 1996; Waters 1990). The same processes that undercut ethnic communities also undercut individual identities. As their worlds become increasingly rootless, defined by a confusing tangle of disparate social networks rather than a cohesive local community, modern individuals long for a sense of self based on something deeper than the culture of the moment (De Vos 1982). They long for the kind of security that they associate with ethnic identities—a sense of self rooted in blood and history, unaltered by the changing styles and shifting affiliations of modern life. In a sense, ethnicity represents an escape from the crisis of identity which accompanies the modernization process. Rather than disappearing, then, ethnicity takes on a different form in the contemporary West. It becomes not a feature of a social group but a feature of the self, a symbol through which individuals construct and express notions of their essential natures. To be Scottish in America does not mean that one has emigrated from Scotland or that one speaks the Scottish language. It means that one regards Scottish origins as somehow intrinsic to one's identity, and that one uses symbols associated with Scottishness to express that identity to oneself and others. This kind of ethnic involvement does not tend to produce effective political organizations or social groups, since it has so little social grounding. Self-described Scots can come from any number of class or occupational groups, and they can imagine Scottishness in a multitude of different ways. As a metaphor for self, on the other hand, symbolic ethnicity can be extremely durable, a fact attested by the growing omnipresence of ethnicity in contemporary western discourse.

Gans's model, while controversial, clearly captures an essential aspect of the development of ethnicity in the United States over the past half century. It also applies very well to the experience of Jewishness among many Copenhagen Jews. Denmark is a self-consciously modern nation, one that has aggressively pursued the kinds of rationalization and democratization associated with the modernization process. Doing so has brought a number of benefits, but it has also brought some of the problems of self-definition characteristic of late modernity. When I interviewed Danish Jews, they often described Jewishness in terms of these problems. They said that Jewishness gave a sort of definition, a distinctiveness that made Jews stand out amid the bland homogeneity of regular Danish society. To be Danish was to be nothing in particular, to be just like everybody else; but Jews had a clear identity, a solid sense of self that was rooted in both blood and history. One of my informants, for example, the daughter of a secularized Jewish man, had taken up the religion and planned to undergo conversion. The reason had nothing to do with any spiritual insight but with a desire for an authentic self. As she put it, "I wanted my life to have boundaries which I had not chosen myself." Such imagery often came up in my interviews, and it seems to describe the kind of attachment many Copenhagen Jews have toward the Jewish community. Even for those who reject Jewish religious beliefs and disdain Jewish ritual practice, the symbols and images of the Jewish tradition offer something valuable—a coherent sense of self amidst the confusion of late modernity.

To summarize, then, and to oversimplify, the past three centuries have seen three broad phases in the external context of the Danish Jewish community. The first, lasting from the community's inception until the beginning of the nineteenth century, involved a strong legal and social separation between Jews and the larger society. In this context, the symbols of the Jewish tradition functioned as the basic framework around which Jewish thought and social life were organized. The second phase began with the Reform period, and lasted until World War II. During this period, the legal boundaries of the Jewish social world were largely dissolved, and the community was formally defined as a religious organization; at the same time, however, both internal networks and external anti-Semitism imposed informal barriers to the full incorporation of Jews into the larger society. In the third phase, beginning after 1945, most of the remaining barriers to Jewish social inclusion fell away. For many Jews, the Jewish tradition became pertinent primarily to personal religious or ethnic identity, rather than to social affiliation or occupational networks. At the same time, however, the growing interest in ethnic identity in the larger society made this dimension of Jewishness particularly appealing. As a result, despite its ebbing significance in structuring Jewish social worlds, the Jewish community has maintained a vital role in the personal experience and self-understanding of individual Jews.

PROSPECTS FOR THE JEWISH FUTURE IN DENMARK

What, then, of the future? In what direction will these contextual trends push the Danish Jewish community over the next century? At least two general patterns seem likely to persist. One is the continuing attenuation of Jewish social ethnicity in Copenhagen. Since the early nineteenth century, those Jewish institutions and networks based in non-religious social connections have steadily eroded. Jewish occupations, for example, have essentially disappeared; likewise, most Jews today share their neighborhoods, their social circles, and increasingly their kinship networks with non-Jews. This process began with the revocation of Jewish legal autonomy in 1814, and it has accelerated since the Danish embrace of Jewry during World War II. We should certainly expect it to continue. The other pattern involves the growing importance of symbolic ethnicity. The cultural trends that make symbolic ethnicity appealing—the decline of local and parochial identities, the fragmentation of class and social networks, the individualization and bureaucratization of modern society—show little evidence of abating. Indeed, they will likely increase as membership in the European Union leads Denmark even further into the globalization process. The symbolic value of Jewish ethnicity, therefore, its ability to offer a sense of rootedness amid the alienation of late modern existence, should become increasingly important to those who hold it.

These trends imply changes for the institutional structure of the Danish Jewish community. As noted above, many of my informants feared that this structure was doomed, that the fragmentation and acculturation of the Danish Jews meant that their institutions would shortly collapse. Certainly some institutions will; those rooted in social ethnicity, relying on closed Jewish social networks for their existence, will almost certainly weaken or fail. The Machsike Hadas synagogue, for example, draws its membership primarily from a specific group of related families. The decline of that group has eroded the synagogue's membership in recent decades; the coming years will likely see it close. Similarly, groups like the Association of Polish Jews, the Jewish Craftsmen's Society, and the Jewish Sewing Club will probably dissolve, along with the cultural and occupational subgroups that created them. At the same time, however, the growing importance of symbolic ethnicity may well strengthen institutions that draw from a wider range of Jewry and that can encompass diverging understandings of Jewishness. The Carolineskole, for example, has managed to attract students from almost every Jewish subgroup, largely because it promises to engage them with Jewish identity. While the school has suffered a number of administrative crises in recent years, its student base has remained strong and it seems poised for a prosperous future. Likewise, WIZO has thrived by offering a setting in which women from a range of backgrounds can discuss their own experiences of Jewishness. Hakoah offers a similarly inclusive setting for men. The interest in Jewish cultural identity has also produced a number of new institutions, like the Jewish museum and a growing list of Jewish cultural journals. To some extent, the government's growing interest in ethnic identity may supplement this trend. In addition to supporting the Jewish nursing homes, for example, the Danish state has recently contributed to the founding of a Jewish museum and several Jewish artistic and cultural projects. European Union policies, which often encourage programs for ethnic and religious minorities, may provide more funding in the future. All of this suggests a more optimistic outlook for Jewish institutions in Copenhagen than the one I heard from many of my informants. The institutional world of Copenhagen Jewry is not collapsing but rather changing, reflecting a change in Jewish identity from a matter of group solidarity to one of individual experience and self-definition. This change opens doors for some institutions even as it closes them for others.

This process has ambiguous implications for the future of Jewish religious practice. On one hand, the decline in the social bases of Jewish identity make Orthodox practice in Copenhagen harder to maintain. As Danish Jews become increasingly diverse and fragmented, the logistical requirements of strict religious observance present serious obstacles. As Jews live farther from the synagogue, for example, they find it harder to come to services, and often virtually impossible to walk to them. Kosher meat becomes harder to find, and it becomes increasingly difficult to find Jewish marriage partners. Beyond such practical issues, Jewish fragmentation erodes the sense of communal solidarity, the palpable feeling of living in a Jewish world, that enlivens Orthodox centers in places like New York and London. As noted in chapter 5, such patterns have caused the orthodox population in Denmark to plummet over the past few decades. In this sense, recent trends in the Jewish community suggest a coming decline in Orthodox religious practice, and presumably a liberal shift for the Mosaiske Troessamfund.

On the other hand, however, symbolic ethnicity implies a powerful role for orthodox observance, among Jews as well as among other ethnic groups. As ethnic identity becomes more individualized, questions of authenticity take on increasing importance for individuals; without a cohesive social group to define it, after all, how does a person establish that his understanding of his primordial ethnicity is valid (Charmé n.d.; Giddens 1991; Waters 1990)? In such settings, religious traditionalists can become important figures in ethnic groups. In their rejection of mainstream secular culture, they seem to embody a deep commitment to the group, making them powerful symbols of ethnic authenticity. They can achieve an influence far beyond their numbers, as acculturated members look to them for models of correct ethnic behavior. To some extent, this process in already taking place among Copenhagen Jews. Most liberal Jews I interviewed described the orthodox as more "really Jewish" than themselves, and they tended to defer to the orthodox in questions of ritual practice (see Buckser 2003). Likewise, many of them evinced a real admiration for Yitzi and Rochel Loewenthal of Chabad House. While they generally knew little about the actual content of Chabad theology, they found the couple's forthright and distinctive Jewishness extremely appealing, and they described them as people who held tightly to their Jewish roots. Indeed, many of those whom I saw at Chabad functions came from a liberal background, attracted less by the Loewenthals' message than by their emphasis on Jewish belonging. Such attitudes suggest a continuing appeal for conservative religiosity, even as its old social base withers away. This appeal may tend to favor a different style of orthodoxy than that of Machsike Hadas, one that emphasizes public visibility for Jewish symbols and a stress on Jewish peoplehood rather than piety. Religious traditionalism in 2050 will likely look very different from its counterpart of 1950. Still, the notion that secularization will sweep away Orthodox Judaism in Copenhagen, that Jews will move irrevocably toward Reform or non-religion, ignores the powerful pull of traditionalism amid the rootlessness of late modernity (see also Lowenthal 1985).

It also ignores one other likely feature of the Jewish future in Denmark: a renewed influx of Jewish immigrants. Waves of immigration have been a recurring feature of Jewish society in Copenhagen since its inception and particularly over the past century. It would be a mistake to overlook the differences among these immigrations—each had its own historical background, and each brought in Jews from a distinctive cultural area. Even so, all of them have shared an origin in the relative tolerance of Danish society, in the fact that Denmark offered a more accommodating environment for Jewish life than many other places in the world. That tolerance seems likely to endure; despite its conflicts over minority assimilation in recent years, Denmark has remained extremely hospitable to Jews, with very little anti-Semitism and a firm commitment to religious freedom. Jewish refugees have been mercifully scarce for the last two decades, of course. State anti-Semitism has declined around the world, and Israel has offered Jews a secure homeland. Jewish history gives little reason, however, to expect such conditions to last indefinitely. And should large numbers of world Jewry find themselves in need of a refuge, we may expect that significant numbers of them will find their way to Denmark. In a sense, the recurrent Jewish immigrations into Denmark reflect not only particular historical circumstances but also an enduring contrast between the relatively hospitable Jewish environment of Denmark and the often dangerous environment of the larger world.

If such immigrations occur, they will almost certainly change the dynamics that have undermined Orthodox practice over the past generation. Given the very high rates of intermarriage and low rates of religious observance among Danish Jewry, almost any immigrants are likely to tilt the community more toward observance and social distinctiveness. This pattern occurred at the opening of the twentieth century, when Eastern European refugees reinvigorated the orthodox wing of what had been an increasingly liberal congregation. It recurred at the close of the century, when immigrants from the Middle East became an important presence in Machsike Hadas and Chabad House. The periodic infusions of Jewish newcomers may therefore pose a counterweight to the social decline of orthodoxy, and their possibility casts into doubt the widespread expectation of the secularization of Danish Jewry. Indeed, they make any definite prediction about the future of the Danish Jewish community chancy at best. If the Jewish community is a field of symbols, then new immigrants make that field broader and richer; they bring with them new symbols of the Jewish tradition and new understandings of established ones. The international nature of the Jewish social world makes it hard to foresee the evolution of any particular community.

What, then, of the question we began with? What will the Jewish community of Denmark look like in 2050, or 2100? The question defies any sort of specific answer. The future of the community depends not only on current trends but on historical events; we cannot know what developments and what travails await the Jews, within and beyond the borders of Denmark. We can, say, however, that it will look like . . . something. The Jewish community is unlikely to disappear, to dissolve into memory through attrition or secularization. Despite all the changes in Danish society over the past century, the body of symbols that makes up the Jewish tradition continues to hold important meanings for thousands of people in Copenhagen. If those meanings are different than they once were, if they now touch more often on symbolic self-definition than on social belonging, they are no less central to personal experience, and they are no more easily discarded. Indeed, as late modern society poses ever greater challenges for individual self-understanding, the seeming primordiality of ethnic identity should make the Jewish community all the more appealing to those who can claim a connection with it. And should the future hold another social transformation in store, whether it be immigration, persecution, religious revival, or something else, the richness of the Jewish symbolic system will offer abundant means by which to understand it.

THE JEWISH FUTURE BEYOND DENMARK

Generalizing this forecast from Copenhagen to the rest of the Jewish world is a risky business. Jewish communities are not interchangeable, and cultural settings that appear similar can conceal deep and important differences. That said, the dynamics of identity, tradition, and cultural context discussed here have striking parallels among Jews elsewhere, particularly in North America, Western Europe, and Australasia. Ethnographies in the United States, for example, have revealed a widespread tendency toward factionalism, with Jewish congregations split over the meaning of Jewish identity and the proper form of religious practice. The opposition between orthodox and liberal Jews in Copenhagen would be entirely familiar to Jews in Muncie (see Rottenberg 1997), Willamette (Zuckerman 1999), or Pittsburgh (Itoh and Plotnicov 1999); so too would controversies over the role of women in services, the requirements of conversion, and the proprieties of kosher observance (see, for example, studies in Belcove-Shalin 1995; Berghahn 1984; Boyarin 1991; Bunzl 2000; Furman 1987; Kugelmass 1986; Kugelmass 1988; Levine 1993; Webber 1994a; Webber 1997). As in Denmark, moreover, these divisions and controversies do not map neatly onto the sociological variables—class, residence, kin group, and so on—that social scientists have normally understood as the basis for community. Jewish identity has become profoundly individualized, with different Jews cobbling together understandings of self and group that accord with their distinctive personal experiences. As Levine has noted regarding New Zealand, "Assimilation, transformation, symbolic ethnicity or ethnic revival do not appear to be mutually exclusive descriptions of the present state of Jewish ethnicity. They appear, at this level at least, to be alternatives produced by the variable interactions of elements of modernity, personal circumstance, family background and the sociocultural resources available to New Zealand Jews" (1993: 342).

Not all Jewish groups echo the Danish situation so strongly, of course; for many Jews, the easy movement between Jewish and non-Jewish circles typical of Copenhagen is all but unthinkable. Ethnographic studies have revealed a broad variation in the circumstances of contemporary Jewish communities. In some settings, barriers of anti-Semitism or political opposition impose sharp boundaries between Jews and other groups (e.g., Adler 2000; Torstrick 2000). In others, the legacy of state anti-Semitism hangs heavily over contemporary Jews, producing a tension between national and Jewish identities very different from that in Denmark (see Chervyakov et al. 1997; Goluboff 2001; Hofman 2000; Mars 1999). In still others, the meaning of Jewishness is deeply affected by associations with an external national identity (see Bettinger-Lopez 2000; Cohen 1999). Such settings present Jews with a very different universe of meanings, a different set of symbols and strategies through which to fashion a Jewish self. Even so, many of them involve some of the same dilemmas of modernity found in Copenhagen, and some of their community dynamics strongly resemble the ones described here. Perhaps most importantly, many of them share the Danish crisis of self-definition, the ongoing tension over national, religious, and ethnic understandings of self and community. As Jews struggle to integrate their various conceptions of Jewishness with their various places in the larger society, they fight to control the religious rituals and social institutions through which their individual resolutions can become manifest. In the process the boundaries of the community become increasingly elusive, even where external barriers would seem to make them self-evident.

The response to such difficulties, among social scientists as well as local Jewish communities, has often been a sense of despair, a feeling that Jewishness stands on the brink of extinction, as the communities that ought to embody it become more and more difficult to pin down. And, indeed, some Jewish populations have dwindled palpably in recent decades, particularly in the former Soviet bloc. Yet for the most part they have not disappeared, and in some areas they have experienced a sort of revival. As in Copenhagen, Jewish identity has persisted in the face of endless confident predictions of its demise. The approach to community outlined here may help us understand why. The variability and disagreement so widespread in Jewish congregations may reflect the modern cultural settings within which they exist; they may portend not the dissolution of the Jewish world but its continuing relevance to an ever more individualized Jewry. If people in Moscow, or Zagreb, or Wellington, or Muncie invoke the Jewish tradition differently from their ancestors, they nonetheless still invoke it, and for many it remains at the heart of their understanding of self. In that sense these Jewish communities endure, and we have no reason to expect them to stop.

The Danish Jews and Conceptions
of Group in Anthropology

I would like to close this discussion with a few words about the more general implications of this analysis for anthropological studies. Much of the difficulty in making sense of the Jewish experience derives from the difficulty of defining the Jews as a bounded group (cf. Schnapper 1987). Being Jewish is not merely an aspect of personal faith or self-perception but an aspect of connection to a collective, a group whose existence is presumed to be associated with a particular set of institutions and ideas. All Jewish Danes whom I interviewed agreed that something like a Jewish community exists, or at least ought to exist. Yet when one tries to put one's finger on the concrete Jewish community, it falls apart, it refuses to be encompassed by any specific description. There is not just one group of Jewish people but many subgroups organized around different and often overlapping criteria. The community is religiously based, yet many in it do not practice a religion; it has a common history, but no one can entirely agree on what that history is or means. The attempt to delineate the specific group of people who belong to the Jewish community, or what they believe or think or do, becomes impossible as soon as one sets out to do it concretely. Our challenge, then, has been to find a way in which we can talk about the Jewish community as something meaningful, yet at the same time encompass the set of transformations and contradictions which Jewish life and social interaction involves.

It is not only Jewish communities that present such a challenge; increasingly, anthropologists are finding similar questions cropping up in studies of virtually all cultures. Early anthropological investigations tended to imagine societies as neatly bounded entities, cultural isolates with integrated systems of social structure, culture, and behavior. Classical community studies focused on populations which seemed to embody this image, islands of culture separated by water or terrain or powerful customs from other groups of people (e.g., Benedict 1934; Malinowski 1927; Mead 1928; Radcliffe-Brown 1948). Even when anthropologists moved to less isolated populations, this paradigm continued to shape their thinking about cultures. When they studied peasant societies, they focused on local networks and cultures rather than regional and international structures (e.g., Pitt-Rivers 1954; Redfield 1956). When they looked at cities, they looked primarily at seemingly isolated underclasses and urban villages (e.g., Belmonte 1979; Gans 1962). In recent decades, however, this notion of soci-

ety has become increasingly untenable. Further research has found powerful external influences even in the apparently isolated fieldsites of classical
anthropology; it turns out to be impossible to understand places like the
Trobriands or Tikopia without reference to the larger cultural and political systems of which they form a part (Barth 1992). Studies of villages have
found residents enmeshed in a variety of regional networks, making the
very definition of a village community problematic (Grönhaug 1978; Han-
nerz 1982). As anthropologists have turned their gaze increasingly toward
modern societies, these difficulties have multiplied. International trade networks, the mass media, cultural tourism, government bureaucracies, and
other features of late modernity make the boundaries between different
groups and cultures difficult to define at all and impossible to define objectively (Appadurai 1996; Giddens 1991; Kearney 1995;Voye 1999;Waters
1990). Individuals have become intimately connected to national and international networks, even as local links among kin and neighbors have
frequently eroded. In such settings, the notion that societies form neatly
structured and separated units, with easily distinguishable cultures, languages, and systems of value, can seem almost quaint. A number of anthropologists have therefore tried to rethink concepts like community,
society, and culture, to find ways of understanding group identities and differences without invoking artificial images of clearly bounded social
groups (e.g.,Appadurai 1996; Kearney 1995; Kuper 1992;Wolf 1982).

Some of these authors have suggested moving away from the old concepts altogether; especially in late modernity, they argue, notions like culture and community no longer have any clear meaning. They have
suggested a focus on the larger-scale systems through which individuals experience the world, the diverse international networks and ethnoscapes that
increasingly define the modern experience (Appadurai 1996; Kuper 1999;
Marcus 1995; Wolf 1982). Such approaches, while revealing, tend to lose
something in the process. While they trace large-scale processes well, they
tend say much less about the sense of local identity that persists even in the
most globalized and media-intensive settings. Even in late modernity—perhaps, indeed, especially in late modernity—individuals base their self-un-
derstandings on their connections to local, occupational, and ethnic
communities (see Bay 1997; Bunzl 2000; Campo 1998; Charmé n.d.; Escobar 1994; Kolsto 1996;Waters 1990).When globalization, bureaucratization,
and other features of modern social organization disrupt such communities,
individuals work all the harder to create new ones; a glance through a large
American newspaper will reveal a multitude of such efforts, ranging from

support groups to Internet communities to communities of retirees or motorcycle riders. Defining distinctive cultures likewise has become a major project for entities ranging from nation-states to ethnic groups to large corporations. We cannot dispense with concepts like culture and community, if only because they have such evident importance to the people we study. Our task, rather, is to find ways of conceptualizing them that accurately reflect the ways in which they actually work.

The analysis of the Jewish community given here certainly does not resolve the problem, but it may suggest a direction for further work. We have suggested seeing a community not as a particular group of individuals, or a particular set of institutions, but as a particular body of symbols, one through which individuals may construct notions of self and connections to other people. Like any set of symbols, a community is subject to a variety of interpretations; questions about the meaning of membership, criteria for belonging, and the relative importance of various features of the community will inevitably vary among individuals and over time. A community's boundaries and functions are not objectively describable but rather vary according to the purposes for which they are invoked. The way the community is understood and enacted, moreover, will vary with its relationship to other symbolic systems through which individuals conceive their worlds. The shape of the Copenhagen Jewish community can be understood only in the context of the larger Danish culture within which it exists; likewise, all communities are influenced by the various regional settings and global networks that shape the lives of their members. This image of the community fits well with the emerging emphasis in anthropology on the changeability and interpenetration of cultures and societies. At the same time, however, it allows us to understand the community's persistent importance to individuals, its role in providing a sense of rootedness and belonging in an ever more dissociated world.

It also suggests lines of comparison in studying the different communities that anthropologists encounter in the contemporary world. The symbolic fields that comprise communities vary in their scope and their content; the details of that variation may explain their relative success in different social and cultural settings. Part of the strength of the Jewish community, for example, is clearly its institutional dimension. Much of Jewish symbolism presupposes an institutional infrastructure, one that includes specific arrangements for praying, for making food, for educating children, for burying the dead, and much more. These institutions play a central part in the enduring vitality of the Danish Jews. As discussed in chapter 4, schools and synagogues and cemeteries make Jewishness something real, providing a concrete place for expressing and contesting the meaning of Jewish identity. They have a particular force in a late modern setting, where tangible evidence of primordial identities has a powerful psychological appeal. Not all communities have a developed institutional dimension, and those that do structure their institutions in a variety of different ways. As we try to understand why some ethnic groups thrive and others founder in contemporary Western Europe, the nature of these institutional structures—the extent to which particular symbolic systems require them, and the forms which they allow them to take—may provide important clues. If we understand communities as bodies of symbols, in other words, the content of those symbols may tell us as much about community survival as statistics about birth rates and language use.

Finally, the experience of the Jews may suggest a more optimistic picture of the future of culture itself. American anthropology has in some ways seemed a tragic discipline for most of its history, having taken culture for its object at the moment of culture's apparent destruction. The image that we fell in love with—the culture of the primitive isolate, self-contained and self-reinforcing, replete with rich and untranslatable local meanings—seemed destined to perish at the hands of the same globalizing forces that brought anthropologists in to study it. Over and over, ethnographies tell of local cultures invaded by colonialism and capitalism, of communities withering in the face of factories and television and Wal-Mart and McDonald's. These little worlds of meaning seem impossibly fragile in the face of such juggernauts, and the whole idea of local culture, of community, can seem inevitably doomed. Yet recent anthropology tells of something else as well, of local resistance to globalizing forces and a stubborn struggle to create community (e.g., Comaroff 1989; Dirks 1994; Hale 1997; Scott 1985). At the margins of the world system, cultures rethink and reinterpret the images of the global marketplace, making even McDonald's into something locally meaningful (cf. Watson 1997). In the heart of modernity, meanwhile, the disconnected and bureaucratized residents of the West look constantly for new bases on which to make something that they can experience as community. American anthropology's understanding of culture may, as some have suggested, have been simplistic and naive. But it does tap into something persistent, some aspect of local experience that seems to matter to people, some desire for community that endures even when modernity says that it shouldn't. And as the Jewish experience suggests, such an impulse can prove surprisingly durable. By many calculations, there shouldn't be a Jewish community in Denmark right now; emancipation should have dissipated it, or secularization should have eviscerated it, or the German occupation should have wiped it out. For that matter, the expulsion of the Jews from Spain, or the destruction of Jerusalem in 132, or the exile of the Jews in Babylon should have ended the story of Jewish culture long ago. Yet it has survived, on the strength of the yearning of its adherents for something it alone can provide. As we watch the denizens of late modernity struggle toward community, despite everything that would seem to make it unattainable, we should never discount the possibility that they may succeed.

Notes

Introduction

1. Coincidentally, there is another well-known pianist and comedian in Denmark named Simon Rosenbaum. The two are not related; in fact, Rosenbaum performs a piece entitled "I Am Not in Any Way Related to Victor Borge, and He Is Not in Any Way Related to Me."

2. It might seem that an easy way out of this dilemma would be to use government statistics on religious affiliation; whatever the different views within the community, the government statistics could stand as an impartial benchmark. No such luck. The figures reported in the Danish government's statistical yearbook are compiled by the official Jewish Community, and they include only members in good standing of the organization. A man who had resigned his membership in the Community would not show up in these figures, though all would agree that he was still a Jew. Members are reported, moreover, by household, not by individual. Since households vary in size, they give only a suggestion of the actual number of member Jews. Other sources of statistics on Jews carry similar problems. After extensive efforts to find a definitive number, I have reluctantly reached the conclusion that no one actually knows how many Jews live in Denmark. The figure of seven thousand is frequently used in conversation and publications, but most of those who use it have no idea where it comes from. (For a discussion of some of the problems associated with the demographics of Jewish populations, see Schnapper 1987).

3. I speak Danish reasonably well, though with a mixture of an American and a West Jutland accent that I suspect many Danes find appalling. When interviewing native Danish Jews, I always conducted interviews in Danish. A number of Jewish immigrants, however, particularly those from the Middle East and Poland, spoke better English than Danish, and a few spoke no

Danish at all. In such cases, I let the informants choose the language of the interview, and they usually chose English.

4. To say that the comments have been helpful does not imply that they have been gentle. Danish Jews who have read my work have tended to find it interesting, and most of them who have spoken with me about it have said that it captured something true about the community. They have also found things to disagree with, however, and they have not been not shy about letting me know. My favorite reaction came from one of the Jewish Community leaders, an academic, who had me over to dinner shortly after the publication of an article in *Ethnic and Racial Studies*. "Andy," he told me, "I have read your latest article, and I have to tell you that it was very interesting and had all the details right, but your thesis is pure anthropological bullshit." Another woman told me that one of my articles in the community journal was "exactly what is wrong with the way people think about Judaism today." I anticipate a rich harvest of comments to follow the present volume.

5. Jack Kugelmass offers an insightful discussion of this issue in the introduction to his edited volume on American Jewry (1988).

Chapter 1

1. A few authors have written general histories of the Jewish community in Denmark; perhaps the best known are Balslev (1932), Borchsenius (1968), and Bamberger (1983). Most of the available historical writing, however, comes in the form of articles or chapters focusing on specific moments in Jewish history, often published in anniversary volumes (e.g., Feigenberg 1984, Margolinsky 1958a, Margolinsky and Meyer 1964). The rescue of the Danish Jews in 1943 has been the subject of extensive historical writing (e.g., Yahil 1969, Flender 1963, Goldberger 1987, Sode-Madsen 1993, Kreth 1995, Buckser 2001, and Paulsson 1995); recently, the immigration period of the early 1900s has also received extensive scholarly attention (Bludnikow 1986).

   I do not know why such a disproportionately large portion of the historical literature on the Danish Jews has been written by authors with last names beginning with B. It is a trend, however, of which I heartily approve.

2. The following characterization of early Jewish history in Denmark is taken primarily from Balslev 1932, Margolinsky 1958a; 1964, Bludnikow and Jørgensen 1984, Borchsenius 1968, Blum 1972, and Bamberger 1983.

3. There are a number of sources for early Jewish population figures in Denmark, not all of which agree. In this chapter I rely mainly on Balslev (1932), who is the source for a number of later histories. While some

sources may give slightly different figures, there is general agreement on the rough shape of the population's growth.

4. See Lausten 1987 for a general account of Danish policies toward nonconformist religions.

5. Population figures here are from Balslev 1932: 20–21.

6. Factionalism in the congregation can also be seen in the proliferation of small burial societies in the early 1700s. See Margolinsky 1958b: 41.

7. The account of the reform period given here is drawn primarily from Linvald 1964, Meyer 1964, Balslev 1932: 29–44, Christensen and Syskind 1984, Bludnikow and Jørgensen 1984, Borchsenius 1968: 61–71, and Gold 1975.

8. This discussion of the Decree of 1814 draws heavily on Gold's analysis in her excellent doctoral dissertation (1975). Other sources used include Margolinsky 1964, Bamberger 1983, Borchsenius 1968, and Balslev 1932. For interested readers, Bamberger 1983: 50–57 provides a full English translation of the decree's text.

9. While our focus here is on the Copenhagen community, it is worth noting that these effects were even more pronounced for Jews living in the provincial towns. Whereas the Copenhagen community was allowed to retain some of its civil administrative functions, as long as they accorded with Danish law, provincial Jews were subjected directly to local civil authorities in all administrative matters.

10. Priest (*præst*) is the standard term for a Danish Lutheran clergyman; the term does not, as in the United States, have a specific association with Catholic clergy. The terminology does matter, however, since the status of a cleric in a state church is considerably different from the traditional status of a rabbi, both in terms of the conditions of employment and of the sacramental responsibilities.

11. This account of the nineteenth century is based primarily on Borchsenius 1968, Bamberger 1983, Balslev 1932, Christensen and Syskind 1984, and Blum 1972.

12. Among the spectators to these events was the fourteen-year-old Hans Christian Andersen, who had just arrived in Copenhagen from outlying Odense. Never having visited the metropolis before, he assumed that such sights were standard in big cities. He relates the story in his autobiography, *Mit Livs Eventyr* (1951).

13. For an excellent discussion of the Mannheimer episode, including an analysis of the liturgical points at issue, see Meyer 1964.

14. The synagogue was installed in the home of Moses Levy in Læderstræde, after Levy had unsuccessfully petitioned the crown to block any reduction of prayers in Wolff's services. The synagogue never received royal approval and was therefore technically illegal, but no effort was ever made to close

it down. The family maintained the synagogue for over a century, holding services twice daily until it closed. Levy himself engaged in a lifelong feud with Wolff; some of his descendants today continue to have antagonistic relations with the Mosaiske Troessamfund. See Bamberger 1983: 63–64.

15. The article appeared without a byline in the community journal *Jødisk Samfund*. Schornstein is generally presumed to be the author, a reasonable presumption which I follow here, but no definite proof of authorship exists.

16. The depiction here of immigrant lives is drawn largely from Bent Bludnikow's excellent study (1986), as well as Welner's evocative memoir of the period (1965). The account of the Machsike Hadas schism draws primarily on Blum 1972 and Melchior 1965.

17. This name is associated with the carrying of myrtle branches during the holiday of Shavuot, and it denotes people who, while not sophisticated in learning, are strong in their faith.

18. The Portuguese minority, of course, formed an important exception; their numbers were always quite limited, however, and their class and occupational position had a great deal in common with the wealthier German Jews.

19. Welner 1965 provides an unusually vivid personal account of this sort of labor and the life circumstances that accompanied it.

20. Welner 1965: 112–113 reports the following conversation between a Jewish immigrant and his elderly father-in-law, who had come to visit him from Poland. The father-in-law had been shocked to see his grandchildren playing with Christian children and their mother drinking coffee in her kitchen with Christian neighbors, and he could not understand the matter-of-fact manner in which these encounters were regarded. Finally, he had to get the mystery answered.

"Gabriel," he said to his stepson, "are the Danes also Christians?"

"Yes, father-in-law," answered Gabriel, "the Danes are not only Christians, they're real Christians."

"And so what are the others?" asked the stepfather.

"The others that you're thinking of, they're *goyim!*"

21. As noted above, there is an extensive historical literature on this period in Danish Jewish history. The most influential among scholars has been Yahil 1969, and accounts like Flender 1963 and Goldberger 1987 have been influential in shaping public perceptions of the event. Among the many other studies, some of the more current and generally oriented include Kreth 1995, Sode-Madsen 1993, Paulsson 1995, Kirchhoff 1995, and Buckser 1998; 1999b; 2001.

22. Logically, this argument would not have protected non-Danish Jews, significant numbers of whom had fled illegally to Denmark just prior to the

occupation, and who had sometimes been turned out of the country if discovered by the police. After the German invasion, however, Jewish refugees were accorded the same protection as Danish Jews, and they were just as likely to be rescued in October 1943.

23. This support came in the highly stylized form of royal protocol; contrary to a widespread legend, the king did not wear a Star of David or threaten to do so. See Lund 1975.

24. The precise mechanics of these arrangements, including their author, remain the subject of considerable unclarity. Best's public account of his actions came in the context of his postwar trial, and he had a strong incentive to minimize his own role in the actions against the Jews. Accounts by different people differed, and historians continue to disagree on exactly what Best did and intended to do in the weeks surrounding the rescue. As these disagreements are mainly of interest to professional historians of the period, I have not reprised them here; interested readers should consult Yahil 1969, Kreth 1995, Paulsson 1995, and Kirchhoff 1995, for discussions of the controversies.

25. Other examples include Melchior 1983, which includes only a few sentences on the postwar period, and Bamberger 1983.

Chapter 2

1. The text is from the sixteenth psalm, and translates roughly as "I keep the Lord always before me." My thanks to David Rokeah for his help.

2. Their tendency to distance themselves from religion does not mean that religious beliefs are absent in Denmark; as Jakob Rod has argued (1972), an underlying set of beliefs about the divine and the supernatural characterizes even many Danes who consider themselves areligious. The basic worldviews of the religious movements that transformed Denmark in the nineteenth century, moreover, continue to have a deep influence on the culture (Buckser 1995; Buckser 1996b; Buckser 2001; Rod 1972).

3. Creating this standard required choosing from many different variations on the interpretation of halakhah, and the process was inevitably a political one. The version that emerged was not an objectively "authentic" traditional Judaism, but rather a negotiated compromise among a number of distinct traditions. For a example of some of the complexities involved in the process, see Fishman 1995.

4. For a discussion of some of these advantages, see Buckser 2003.

5. Books in English about the Danish Jews routinely use this term to refer to the MT synagogue, and I follow that practice here. In conversations with Danish Jews, however, the building is usually just called "the synagogue" (*synagogen*); one might call it "the big synagogue" (*den store synagog*), but

only when distinguishing it from one of the small synagogues. "The Great Synagogue" is a defensible translation of "den store synagog," but it suggests a grandeur not present in the Danish, and it is in no sense a formal name for the building.

6. This rule derives from the prohibition against working on the Sabbath; carrying something is considered labor. The only exception made is for housekeys, which members may bring with them into the synagogue. Violating the rule would be quite easy, since the guards do not physically check those coming in, but most Jews whom I spoke with at the services— even quite liberal ones—said that they made a point of not carrying anything with them.

7. These caps are always referred to by the Hebrew term *kippah* (pl. *kippot*); none of my interview subjects used the Yiddish term, *yarmulke,* which is commonly used in the United States.

8. The service is often called the *Kol Nidre,* in reference to the prayer that opens the service. In this prayer, Jews ask God to release them from any vows that they have taken during the year but failed to keep. The service includes many other prayers as well.

9. Kosher laws mandate the strict separation of milk and meat.

10. Unbeknownst to myself, I was breaking not only Jewish eating rules; I was also eating something that a respectable academic shouldn't have. Two years later, I mentioned to a Copenhagen anthropologist that I loved the combination of Danish hot dogs and chocolate milk, and he looked at me with an expression of shock and disbelief. "You can't be serious!" he said. "That's working-class food. That's what truck drivers eat! You see them pulled off the road, eating their hot dogs with chocolate milk. You mustn't eat that!" He then told me at length what was appropriate for me to eat, advice that I have made a point of ignoring. The Danes really do make superb chocolate milk.

11. Technically, the terms *bar mitzvah* and *bat mitzvah* do not refer to the ceremonies but to the subjects of them. A boy becomes *bar mitzvah,* literally a "son of the commandments," when he turns thirteen, regardless of whether any ceremony is carried out; a girl becomes *bat mitzvah,* "daughter of the commandments," when she turns twelve. At that point, according to *halakhah,* they become subject to the ritual requirements of adult Jews. In practice, however, most Jewish congregations hold some sort of ceremony to mark the occasion, particularly for boys, and these ceremonies are commonly referred to as *bar* or *bat mitzvahs.*

12. She told me this during a group discussion at a meeting of a Jewish youth organization. The leader of the meeting confirmed the point, but said that budgetary constraints might force the Jewish Community to change this policy. Then, with a grin, he uttered the most effective laugh line I heard during my fieldwork: "Jo, også med omskæringer skal der være ned-

skæringer" ("Well, even with cutoffs there have to be cutbacks"). It took a minute or so to restore order.

13. In addition to rules specifically laid out in the scriptures, a number of practices have developed in Jewish culture that are generally understood as religious imperatives, even though they have no explicit basis in scripture. The common practice for men of wearing a headcovering, for example, has no scriptural basis; nonetheless, it is understood by most Jews as a religious requirement, and going without a *kippah* is generally regarded as a failure of observance. In this section, I do not distinguish between such generally accepted non-scriptural religious rules and those actually set out in the sacred texts.

14. For a more extended ethnographic and theoretical treatment of this subject, see Buckser 1999a.

15. A third kosher shop recently opened in Frederiksberg, aiming to draw on the relatively large concentration of Jews in that area of the city. Whether it will survive is an open question; while the proprieters hope that it can also draw business from Middle Eastern immigrants, most of those I asked about it gave it little chance for survival. It is perhaps worth noting that the main proprietor of this business is not Jewish, nor is the operator of the Kosher Delicatessen.

16. One of the delicatessens, the Kosher Delikatessen in Lyngbygade, requires ritual assistance and official patronage from the chief rabbi to stay in operation; the other, Samson Kosher, uses a thriving cheese export business to subsidize its local grocery services.

17. "Family" here includes unmarried couples living together. Such arrangements are very common in Denmark, and they do not carry a moral stigma; couples may live together for decades without being married.

Chapter 3

1. In contrast to many Jewish communities, academics hold a relatively low profile in Jewish Copenhagen. There are relatively few of them, and not many play important roles in the MT. A number of Jewish intellectuals lamented the situation in interviews, with one describing Copenhagen as the most *petit bourgeois* Jewish community he had ever seen.

2. *Hygge* is a sense of coziness and security experienced in settings of intimacy and acceptance. It does not translate well into English, and American culture has no real equivalent of the concept.

3. This was a matter of social courtesy—I am in fact quite big, but am by no stretch of the imagination handsome.

4. I answered this question, as I sometimes did in my fieldwork, somewhat obliquely; rather than saying yes or no, I said, "My father was, but not my

mother." Most Danish Jews responded to this formulation by saying, "So you're not Jewish." At the Klub, by contrast, the immediate response was, "Then you're Jewish!"

5. The historical material presented here draws on interviews with Polish Jewish immigrants, as well as from Nina Roth's excellent article (1999).

6. I should emphasize that this attribution has no connection with the American ethnic slur that typecasts Poles as unintelligent. The latter stereotype does not occur in Denmark, where Poles, if stereotyped, are typecast as obsessed with honor. When Danes tell jokes about people of low intelligence, their standard target is the residents of an area of Jutland known as Mols. The derogatory references to the Polish Jews are based on their difficulties in mastering the Danish language. It is also worth pointing out that these stereotypes are singularly inaccurate; the Polish Jewish immigrants from 1969 had an unusually high level of education, and they include a number of quite distinguished scientists.

7. In an article on the Polish Jews published in a Jewish journal in 1982, Jacques Blum writes, "[Many Polish Jews] inform me that the *worst* form of prejudice and discrimination that they have experienced, whether in Poland or in Denmark, has come from certain circles within the Mosaiske Troessamfund."

8. Guttermann's prominence in disputes between liberals and the orthodox in Copenhagen has made him reviled among large parts of the congregation. Since my early interviews were largely among liberal Jews, I initially imagined him as something of an ogre; informants described him as a narrow-minded zealot and doubted whether he would even speak to me. When he did agree to an interview, at the Villa Strand, I dreaded the encounter for days in advance. I was shocked, on meeting him, to find a very small, thin, cheerful man, with a ready wit and a willingness to talk about anything I asked him. Later I learned that he drives a red convertible and has been the subject of a television profile in a series on offbeat characters.

Chapter 4

1. The spelling of this title varies with the context, which can be confusing for English readers. *Troessamfund* is an archaic spelling of *trossamfund*, literally "community [*samfund*] of faith [*tro*]." *Mosaisk* translates most directly as "Jewish," but it implies an association with the religious aspect of Jewish identity, as opposed to the more ethnic connotations of *jødisk*. The two words are often preceded by the definite article "det," producing *Det Mosaiske Troessamfund*. The name need not always take an article, however, which affects the spelling—Danish adjectives add an *e* when preceded by a definite article. Moreover, unlike English, Danish does not automatically

capitalize adjectives referring to religious faiths. As a result, readers may encounter *Mosaiske, mosaiske, Mosaisk,* or *mosaisk* when reading about the community. All mean the same thing, as do *troessamfund* and *trossamfund.*

2. Danes use acronyms very frequently, much more than Americans. Where American newspapers generally refer to Republicans and Democrats, for example, Danish newspapers refer to most political parties by their initials. The Mosaiske Troessamfund, likewise, is usually shortened to MT, both in writing and in speech.

   This tendency also applies, incidentally, to personal names, which for well-known people very often take the form of the first two initials plus the last name. Thus Danes refer to theologian N. F. S. Grundtvig, statesman A. P. Bernstorff, MT founder M. L. Nathanson, and fairy tale author H. C. Andersen. The latter usage can cause confusion for Americans. When naming individual letters, Danes pronounce H like the English "hoe" and "C" like the English "say." A friend of mine from California, on his first visit to Denmark, spent several weeks wondering who José Andersen might be.

3. In Denmark, records of births, deaths, and marriages are recorded and reported by parish priests in the state church. Officially recognized alternative religious groups (*trossamfund*) maintain records for their members. The MT is such a group, and accordingly the rabbi maintains two sets of *ministerialbøger,* the massive record books in which demographic events are recorded.

4. Danes who pay dues to a community like the MT are exempt from the church tax. If a Jew leaves the MT, however, even without joining the state church, he or she must resume payment of the church tax.

5. New Outlook was founded with an English name, and still uses one; its journal title, however, uses the Danish term for outlook, *Udsyn.*

6. For a list of major voluntary associations with a connection to the MT, see Dessauer et al. (1996: 36–41). This list, presented in an official MT publication in 1996, is not complete, but it does detail the origins and purposes of many of the associated groups.

7. *Kirkegaard* means, literally, "churchyard;" it is the standard word for cemetery in Denmark, where country churches have their cemeteries immediately surrounding them. Some Jews object to the intrusion of the word *kirk* in this context, preferring the alternative term *begravelsesplads,* or "burial place." Most Jews I spoke with, however, used *kirkegaard,* and so do most publications, including some produced by the MT.

8. Assessments of the academic quality of the school fell into two camps. A number of my interview subjects, especially those with children in the school, described the Carolineskole as one of Copenhagen's premier educational institutions. They said that the students scored very well on

exams, and they pointed to the high percentage of its graduates who go on to *gymnasium*. Others criticized the school quite harshly, comparing it unfavorably with Jewish schools in other countries. The high gymnasium matriculation, they said, results not from the quality of the school's instruction but from its location and the class background of the families who can afford its tuition. I have no basis on which to favor either of these views. It may be noteworthy, however, that parents almost always told me that they had sent their children to the school because of concerns about identity; they very seldom said that they had chosen it for its academic quality.

9. This characterization of the events of 1981–1982 is based on interviews with several of the principals, as well as press accounts from the time. It is worth noting that while informants differed in their versions of what occurred, they concurred in the major features of the conflict, including the centrality of the personal feud between Melchior and Guttermann.

10. This campaigning, according to some sources, included attempts to intimidate potential opponents. One man, for example, told me of vague but fairly aggressive threats that Melchior had made to him over the telephone when he was considering supporting Rudaizky. While I cannot verify such accounts, my impression from a variety of interviews, with both supporters and opponents of the rabbi, is that Melchior pursued his cause energetically and, in some cases, somewhat vindictively.

11. This letter was published in the community journal, *Jødisk Orientering*, which said that the typewriter used to write it had been matched to the one Svend Meyer used for the Action Committee. Melchior subsequently sent a letter to the Polish voluntary associations, distancing himself from the sentiments and conveying Meyer's own apologies.

12. Text translated from photostat of letter published in *Jødisk Orientering*, October 1981; my translation.

Chapter 5

1. As in much of Europe, the political left in Denmark tends to be highly critical of Israel, in ways which sometimes have anti-Semitic overtones.

2. Both groups, for example, use similar methods of ritual slaughter to produce meat. Animal rights groups have worked to abolish these methods as an unnecessary cruelty, and they have been supported by right-wing politicians opposed to Muslim immigrants. Sweden banned ritual slaughter in the 1990s; in 1996, Jews and Muslims worked together to turn back a similar proposal in Denmark. Jews and Muslims also have similar interests in the promotion of religious and ethnic tolerance, and at times leaders have worked together on such issues.

Chapter 6

1. The lyrics are, of course, a bit more complex than they sound. The appeal of the song is partly a comic one, involving a contrast between the style of the music and the sentiment expressed; the band's image was one of a loud rock band that was also somewhat cuddly. One should not get the impression that Danish teenagers were shouting lyrics about pudding without any ironic meaning. Nonetheless, the song's referent indicates the centrality of Danishness to the discourse of popular culture. Even the reference to rhubarb pudding carries meaning. It is, on the one hand, something of a national dessert, a thick sweet sauce that, when served with heavy cream, makes Danish palates tingle and makes heart surgeons get out their instruments. On the other hand, the name of the pudding is a notoriously difficult set of words to pronounce, and virtually no one but a native Dane can say it convincingly. Singing the song not only restates a classic piece of Danish culinary culture; it also restates the singer's mastery of a difficult and little-known language.

2. One interesting exception to this was the soccer game I observed (see chapter 4). The Jewish team's opponents were very demonstrative in their play, particularly when disappointed; a bad play would often be accompanied by a loud shout of disgust or cursing from the player who had made it. The Jews, by contrast, were quiet and reserved, with only one player—the goalie—cursing or yelling at his teammates.

3. The largest exception to this general trend was the Polish Jews, who were invariably warm and inviting when I interviewed them. They themselves, however, had often felt similarly excluded by the native Danish Jews.

4. Contrary to American stereotypes, Danes are not all blonds; indeed, on the island where I did my initial fieldwork, dark brown was perhaps the most common hair color. One sees more blonds on the streets of Copenhagen, and I once speculated to my Jutland neighbor that more people there must be intermarried with Swedes. He laughed, and said that it had nothing to do with Swedes but rather with hair dye.

Chapter 7

1. This very abbreviated discussion of the rescue leaves out, of necessity, both much of the historical complexity of the rescue and many of its most interesting stories. Readers interested in exploring the rescue further may turn to a number of comprehensive studies, as well as many articles on specific aspects of the events. For English speakers, the two best available works are Yahil 1969 and Goldberger 1987; for those who read Danish, several excellent and much more recent studies are available (e.g., Kreth 1995; Sode-Madsen 1993). Readers may also wish to consult my own articles on the

rescue for a further discussion of some of the themes explored here (Buckser 1998; Buckser 1999b; Buckser 2001).

2. For a thorough treatment of this issue, see Lund 1975.

3. In 1942, Christian sparked a major rift with Germany by sending an terse reply to a flowery birthday telegram from Adolf Hitler; the ensuing "Telegram Crisis" was a major ingredient in the collapse of the Buhl government and the installation of Erik Scavenius as premier. For a discussion of the political situation of the time, see Jones 1970: 166–181.

4. The precise details of this operation are shrouded in mystery and are an ongoing subject of debate among historians. I have done my best to skirt such issues in this treatment, describing events in a way that accords with the general historical consensus without venturing into unsettled details. Readers interested in such questions, however, may find extensive treatments in Yahil 1969, Sode-Madsen 1993, Goldberger 1987, Kirchhoff 1995, and Paulsson 1995.

5. For a discussion of the role of doctors and hospitals in the rescue, see Svendstorp 1946.

6. For two particularly evocative accounts of the experience of Danish Jews in Theresienstadt, see M. Oppenhejm 1981 and R. Oppenhejm 1966. The MT has also published a brief volume (Sode-Madsen 1995) with a variety of materials on the experience of internees.

7. The visit was a transparent charade, in which the camp was cleaned and beautified to give the impression that the prisoners were well-treated. In a particularly cruel act, the authorities shipped a large number of prisoners to the death camps, so that Theresienstadt would not appear overcrowded. In later years, a number of people have criticized the Danish Red Cross for participating in the event, which not only broadcast German propaganda but sent thousands of prisoners to their deaths. In its defense, however, the Danish delegation was well aware of the deception; its concern was to establish the physical well-being of the Danish internees, which it was able to do. Those sent to the death camps, moreover, would likely have been sent there anyway, as most residents of Theresienstadt ultimately were.

8. Ironically, while the Jews were hailed as heroes during their passage through Denmark, their arrival in Sweden came with absolutely no fanfare. Unsure of just when they would arrive, Jewish leaders had made no arrangements for an organized reception; as a result, when they finally reached safety after their dramatic passage, they found only Rabbi Melchior and Louis Meyer standing on the dock to greet them. Some of them were irritated at the contrast between their jubilant reception by Danish Christians and their seeming neglect by Danish Jews. An account of the event can be found in Melchior's memoirs (Melchior 1965).

9. The names and identities of informants have been altered to protect their privacy, as have some place names and sequences of events.

10. See, for example, Ardener 1989, Badone 1991, Bloch 1977 Buckser 1995; 1998; 1999b, Errington 1996, Friedman 1992, Handler 1984, Herzfeld 1982; 1991, Hobsbawm 1984, Lowenthal 1985; 1996, Mitchell 1998, Trevor-Roper 1984, and Zerubavel 1994.

Bibliography

Adler, Franklin Hugh
  2000   South African Jews and apartheid. Patterns of Prejudice 34(4):23–36.
Alba, Richard D.
  1985   The twilight of ethnicity among Americans of European ancestry: the
          case of Italians. Ethnic and Racial Studies 8(1):134–158.
  ——
  1990   Ethnic Identity: The Transformation of White America. New Haven: Yale
          University Press.
Andersen, Hans Christian
  1951   Mit Livs Eventyr. Copenhagen: Gyldendal.
Anonymous
  1981   Nyt fra Repræsentanterne. In Jødisk Orientering. Pp. 18, Vol. 52.
Appadurai, Arjun
  1990   Disjuncture and Difference in the Global Cultural Economy. In Global
          Culture: Nationalism, Globalization and Modernity. M. Featherstone, ed.
          Pp. 295–310. London: SAGE Publications.
  ······
  1996   Modernity at Large: Cultural Dimensions of Globalization. Minneapo-
          lis: University of Minnesota Press.
Ardener, E.
  1989   The construction of history: "Vestiges of creation." In History and Eth-
          nicity. E. Tonkin, M. McDonald, and M. Chapman, eds. Pp. 22–23. New
          York.
Arnheim, Arthur
  1950   Jøderne i Danmark—statistisk set. Jødisk Samfund 24(3):18–19.
Badone, Ellen
  1991   Ethnography, fiction, and the meanings of the past in Brittany. American
          Ethnologist 18(3):518–545.
Balle-Petersen, M.
  1983   Grundtvigske kulturmiljøer. In Efterklange: Et Grundtvig-Seminar. J.
          Jensen and E. Nielsen, eds. Pp. 64–79. Aarhus: Aarhus University Press.

Balslev, Benjamin
1932 De danske Jøders historie. Copenhagen: Lohse.
Bamberger, Ib Nathan
1983 The Viking Jews: A History of the Jews of Denmark. New York: Shengold.
Barfod, Jørgen H., Norman L. Kleeblatt, and Vivian B. Mann, eds.
1983 Kings and Citizens: The History of the Jews in Denmark 1622–1983. New York: The Jewish Museum.
Barth, Fredrik
1992 Toward greater naturalism in conceptualizing societies. In Conceptualizing Society. A. Kuper, ed. Pp. 17–33. London: Routledge.
Bay, Joi
1997 Fra ungdoms-kultur til subkultur: rockere og bikere. Dansk Sociologi 8(3):7–25.
Begtrup, Holger
1936 N. F. S. Grundtvigs Danske Kristendom. Copenhagen: G. E. C. Gads.
Belcove-Shalin, Janet S., ed.
1995 New World Hasidim: Ethnographic Studies of Hasidic Jews in America. Albany: State University of New York Press.
Belmonte, Thomas
1979 The Broken Fountain. New York: Columbia University Press.
Benedict, Ruth
1934 Patterns of Culture. Boston: Houghton-Mifflin.
Ben-Sasson, H. H.
1976 A History of the Jewish People. Cambridge: Harvard University Press.
Berghahn, Marion
1984 German-Jewish Refugees in England: The Ambiguities of Assimilation. New York: Macmillan.
Bettinger-Lopez, Caroline
2000 Cuban-Jewish Journeys: Searching for Identity, Home, and History in Miami. Knoxville: University of Tennessee Press.
Bhatt, Chetan
2000 Dharmo Rakshati Rakshitah: Hindutva movements in the U.K. Ethnic and Racial Studies 23(3):559–593.
Blau, Joseph L.
1966 Modern varieties of Judaism. New York: Columbia University Press.
Bloch, Maurice
1977 The past and the present in the present. Man 12(3):278–292.
Blüdnikow, Bent
1986 Immigranter: De østeuropæiske jøder i København 1905–1920. Valby, Denmark: Borgen.

———
1991 Som om de slet ikke eksisterede: Hugo Rothenberg og kampen for de tyske jøder. Copenhagen: Samleren.

Blüdnikow, Bent, and Harald Jørgensen
1984   Den lange vandring til borgerlig ligestilling i 1814. *In* Indenfor Murene: Jødisk liv i Danmark 1684–1984. M. Feigenberg, ed. Pp. 13–90. Copenhagen: C. A. Reitzel.

Blum, Jacques
1972   Danske og/eller Jøde: En kultursociologisk undersøgelse af den jødiske minoritet i Danmark. Copenhagen: Gyldendal.

——

1982   Indvandrere og minoriteter: fordomme og diskrimination i det danske samfund. Copenhagen: Gad.

——

1986   Splinten i øjet: Om danskernes forhold til de fremmede. Odense: Stavnsager.

Borchsenius, Poul
1968   Historien om de Danske Jøder. Copenhagen: Fremad.

Boyarin, Jonathan
1991   Polish Jews in Paris: The Ethnography of Memory. Bloomington: Indiana University Press.

——

1992   Storm From Paradise: The Politics of Jewish Memory. Minneapolis: University of Minnesota Press.

——

1996   Thinking in Jewish. Chicago: University of Chicago Press.

Buckser, Andrew
1995   Tradition, power, and allegory: Constructions of the past in two Danish religious movements. Ethnology 34:257–272.

——

1996a   Communities of Faith: Sectarianism, Identity, and Social Change on a Danish Island. Providence, RI: Berghahn.

——

1996b   Religion, science, and secularization theory on a Danish island. Journal for the Scientific Study of Religion 35(4):432–441.

——

1998   Group identities and the 1943 rescue of the Danish Jews. Ethnology 37(3):209–226.

——

1999a   Keeping kosher: Eating and social identity among the Jews of Denmark. Ethnology 38(3):191–209.

——

1999b   Modern identities and the creation of history: Stories of rescue among the Jews of Denmark. Anthropological Quarterly 72(1):1–17.

——

2000   Jewish identity and the meaning of community in contemporary Denmark. Ethnic and Racial Studies 23(4).

2001   Rescue and cultural context during the Holocaust: Grundtvigian nationalism and the rescue of the Danish Jews. Shofar 19(2):1–25.

2003   Religious practice and cultural politics among the Jews of Copenhagen. American Ethnologist 30(1).

Bunzl, Matti
    2000   Resistive play: Sports and the emergence of Jewish visibility in contemporary Vienna. Journal of Sport & Social Issues 24(3):232–250.

Campo, Juan Eduardo
    1998   American pilgrimage landscapes. Annals of the American Academy of Political Science 55(8):40–56.

Charmé, Stuart Z.
    n.d.   Who is a "real" Jew?: Models of authenticity in contemporary Jewish identity.

Chervyakov, Valeriy, Zvi Gitelman, and Vladimir Shapiro
    1997   Religion and ethnicity: Judaism in the ethnic consciousness of contemporary Russian Jews. Ethnic & Racial Studies 20(2):280–305.

Christensen, Lorenz
    1943   Det tredie ting. Copenhagen.

Christensen, Merete, and Britta Syskind
    1984   De danske joders livsvilkår 1814–1905. In Indenfor Murene: Jødisk liv i Danmark 1684–1984. M. Feigenberg, ed. Pp. 91–142. Copenhagen: C. A. Reitzel.

Cohen, Anthony
    1985   The Symbolic Construction of Community. London: Tavistock.

Cohen, Rina
    1999   From ethnonational enclave to diasporic community: The mainstreaming of Israeli Jewish migrants in Toronto. Diaspora 8(2):121–136.

Comaroff, John L.
    1989   Images of empire, contests of conscience: Models of colonial domination in South Africa. American Ethnologist 16(4):661–685.

Cuddihy, Johon Murray
    1987   The Ordeal of Civility: Freud, Marx, Levi-Strauss, and the Jewish Struggle with Modernity. Boston: Beacon.

De Vos, George
    1982   Ethnic pluralism: Conflict and accommodation. In Ethnic identity: Cultural continuities and change. G. D. Vos and L. Romanucci-Ross, eds. Chicago: University of Chicago Press.

Della Pergola, Sergio
    1991   New data on demography and identification among Jews in the U.S.: Trends, inconsistencies and disagreements. Contemporary Jewry 12:67–97.

Dessauer, John, Ino Jacobsen, and Bent Melchior, eds.
    1996   Det Mosaiske Troessamfund: Det Jødiske samfund i Danmark. Copenhagen: Det Mosaiske Troessamfund.

Dirks, Nicholas
   1994   Ritual and resistance: Subversion as social fact. *In* Culture/Power/History: A reader in contemporary social theory. N. B. Dirks, G. Eley, and S. B. Ortner, eds. Pp. 483–503. Princeton: Princeton University Press.
Dreidger, Leo
   1980   Jewish identity: the maintenance of urban religious and ethnic boundaries. Ethnic and Racial Studies 3(1):67–88.
Elazar, Daniel, Edina Weiss Liberlies, and Simcha Werner, eds.
   1984   The Jewish Communities of Scandinavia: Sweden, Denmark, Norway, and Finland. Lanham, MD: University Press of America.
Enoch, Yael
   1994   The intolerance of a tolerant people: Ethnic relations in Denmark. Ethnic and Racial Studies 17(2):282–300.
Errington, Frederick, and Deborah Gewertz
   1996   The invention of tradition in a Papua New Guinean modernity. American Ethnologist 21(1):104–122.
Escobar, Arturo
   1994   Welcome to Cyberia. Current Anthropology 35(3):211–232.
Feigenberg, Meier, ed.
   1984   Indenfor murene: Jødisk liv i Danmark 1684–1984. Copenhagen: C. A. Reitzel.
Finkielkraut, Alain
   1994   The imaginary Jew. K. O'Neill and D. Suchoff, transl. Lincoln, NE: University of Nebraska Press.
Fishman, Aryei
   1995   Modern Orthodox Judaism: A study in ambivalence. Social Compass 42(1):89–95.
Flender, Harold
   1963   Rescue in Denmark. New York: Holocaust Library.
Friedman, Jonathan
   1992   The past in the future: History and the politics of identity. American Anthropologist 94(4):837–859.
Furman, Frida Kerner
   1987   Beyond Yiddishkeit : The Struggle for Jewish Identity in a Reform Synagogue. Albany: State University of New York Press.
Gans, Herbert
   1962   The Urban Villagers: Group and Class in the Life of Italian-Americans. Glencoe, Ill.: Free Press.
   ——
   1979   Symbolic ethnicity: The future of ethnic groups and cultures in America. Ethnic and Racial Studies 2(1):837–859.
Gans, Herbert J.
   1994   Symbolic ethnicity and symbolic religiosity: Towards a comparison of ethnic and religious acculturation. Ethnic and Racial Studies 17(4):577–592.

Geertz, Clifford
1973   The Interpretation of Cultures. New York: Basic Books.

———

1988   Works and Lives: The Anthropologist as Author. Palo Alto: Stanford University Press.
Giddens, Anthony
1991   Modernity and Self-identity: Self and Society in the Late Modern Age. Cambridge: Polity.
Gilman, Sander L.
1986   Jewish Self-Hatred: Anti-Semitism and the Hidden Language of the Jews. Baltimore: Johns Hopkins University Press.
Glazier, Stephen D.
1996   New World African ritual: Genuine and spurious. Journal for the Scientific Study of Religion 35(4):420–431.
Gold, Carol
1975   The Danish Reform Era, University of Wisconsin.
Goldberg, Harvey, ed.
1987   Judaism Viewed from Within and from Without: Anthropological Studies. Albany: State University of New York Press.
Goldberger, Leo, ed.
1987   The Rescue of the Danish Jews: Moral Courage under Stress. New York: New York University Press.
Golden, D.
2001   Storytelling the Future: Israelis, Immigrants and the Imagining of Community. Anthropological Quarterly 75(1):7–35.
Goldscheider, Calvin, and Alan S. Zuckerman
1984   The Transformation of the Jews. Chicago: University of Chicago Press.
Goldschmidt, Meir
1867   Ravnen. Copenhagen: C. Steen.

———

1968   En Jøde. Copenhagen: Gyldendal.
Goluboff, Sascha L.
2001   Fistfights at the Moscow Choral Synagogue: Ethnicity and ritual in post-Soviet Russia. Anthropological Quarterly 74(2):55–71.
Gullestad, Marianne
1989   Small facts and large issues: The anthropology of contemporary Scandinavian society. Annual Review of Anthropology 18:71–93.

———

1994   Sticking together or standing out: A Scandinavian life story. Cultural Studies 8(2):253–268.
Hale, Lindsay Lauren
1997   Preto velho: Resistance, redemption, and engendered representations of slavery in a Brazilian possession-trance religion. American Ethnologist 24(2):392–414.

Hamberg, Eva, and Thorleif Pettersson
1994 The religious market: Denominational competition and religious participation in contemporary Sweden. Journal for the Scientific Study of Religion 33:205–216.

Handler, Richard, and Jocelyn Linnekin
1984 Tradition, genuine or spurious. Journal of American Folklore
97(385):273–290.

Herzfeld, Michael
1982 Ours Once More: Folklore, Ideology, and the Making of Modern
Greece. Austin: University of Texas Press.

—— 1991 A Place in History: Social and Monumental Time in a Cretan Town.
Princeton, NJ: Princeton University Press.

Hobsbawm, Eric, and Terence Ranger, eds.
1984 The Invention of Tradition. Cambridge: Cambridge University Press.

Hofman, Nila Ginger
2000 The Jewish Community of Zagreb : Negotiating Identity in the New
Eastern Europe. Doctoral dissertation, Purdue University.

Hunt, Lynn, ed.
1996 The French Revolution and Human Rights: A Brief Documentary History. New York: St. Martin's Press.

Itoh, Reiko, and Leonard Plotnicov
1999 The Saturday morning informal service: Community and identity in a
Reform congregation. Ethnology 38(1):1–19.

Jenkins, Richard
1996 Ethnicity etcetera: Social anthropological points of view. Ethnic and
Racial Studies 19(4):807–822.

Jokinen, Kimmo
1994 Cultural uniformity, differentiation, and small national cultures. Cultural
Studies 8(2):208–219.

Jones, W. Glyn
1970 Denmark. London: Praeger.

Kaufmann, Hanne
1970 Alle disse skæbner. Copenhagen: Branner & Koch.

Kearney, M.
1995 The local and the global: The anthropology of globalization and transnationalism. Annual Review of Anthropology 24:247–65.

Kirchhoff, Hans
1995 Denmark: A light in the darkness of the Holocaust? A reply to Gunnar
S. Paulsson. Journal of Contemporary History 30:465–479.

Kivisto, Peter, and Ben Nefzger
1993 Symbolic ethnicity and American Jews: The relationship of ethnic
identity to behavior and group affiliation. Social Science Journal
30(1):1–12.

Knudsen, Anne
1996   Her går det godt, send flere penge. Copenhagen: Gyldendal.
Kolsto, Pal
1996   The new Russian diaspora—an identity of its own? Ethnic and Racial
        Studies 19(3):609–39.
Kreth, Rasmus
1995   Flugten til Sverige: Aktionen mod de danske jøder oktober 1943.
        Copenhagen: Gyldendal.
Kugelmass, Jack
1986   The Miracle of Intervale Avenue : The Story of a Jewish Congregation
        in the South Bronx. New York: Schocken.
—, ed.
1988   Between Two Worlds : Ethnographic Essays on American Jewry. Ithaca:
        Cornell University Press.
Kuper, Adam
1988   The Invention of Primitive Society. London: Routledge.
—, ed.
1992   Conceptualizing Society. London: Routledge.
—
1999   Culture:The Anthropologist's Account. Cambridge: Harvard University Press.
Lake, Obiagele
1995   Toward a Pan-African identity: Diaspora African repatriates in Ghana.
        Anthropological Quarterly 68(1):21–36.
Lausten, Martin Schwarz
1987   Danmarks Kirkehistorie. Copenhagen: Gyldendal.
Lazerwitz, Bernard, et al.
1998   Jewish choices: American Jewish denominationalism. Albany: State Uni-
        versity of New York Press.
Levine, H. B.
1993   Making sense of Jewish ethnicity: Identification patterns of New Zealan-
        ders of mixed parentage. Ethnic and Racial Studies 6(2):323–344.
Lindhardt, P. G.
1959   Vækkelse og kirkelige retninger. Copenhagen: Hans Rietzel.
Linvald, Steffen
1964   Den jødiske frihedskamp i Danmark og Anordningen af 29. Marts 1814.
        In Ved 150 aars-dagen for Anordningen af 29. Marts 1814. J. Margolinsky
        and P. Meyer, eds. Pp. 11–34. Copenhagen: Det Mosaiske Troessamfund.
Louie, Andrea
2000   Re-territorializing transnationalism: Chinese Americans and the Chi-
        nese motherland. American Ethnologist 27(3):645–669.
Lowenthal, David
1985   The Past is a Foreign Country. Cambridge: Cambridge University Press.
—
1996   Possessed by the Past: The Heritage Crusade and the Spoils of History.
        New York: Free Press.

Lund, Jens
    1975   The legend of the king and the star. Indiana Folklore 8:1–2.
Malinowski, Bronislaw
    1927   Sex and Repression in Savage Society. London: Routledge & Kegan
            Paul.
Malkki, Liisa
    1995   Refugees and exile: From refugee studies to the national order of things.
            Annual Review of Anthropology 24(495–523).
Marcus, George E.
    1995   Ethnography in/of the world system: the emergence of multi-sided
            ethnography. Annual Review of Anthropology 24:95–117.
Margolinsky, Julius, ed.
    1958a   Chevra Kaddischa 1858–1958. Copenhagen: Det Forenede Israelitiske
            Begravelsesselskab.

    ———

    1958b   Træk af chevra kaddischa's historie. In Chevra Kaddischa 1858–1958. J.
            Margolinsky, ed. Pp. 32–112. Copenhagen: Det Forenede Israelitiske Be-
            gravelsesselskab.
Margolinsky, Julius, and Poul Meyer, eds.
    1964   Ved 150-aars-dagen for Anordningen af 29. Marts 1814. Copenhagen:
            Det Mosaiske Troessamfund.
Mars, Leonard
    1999   Discontinuity, tradition, and innovation: Anthropological reflections on
            Jewish identity in contemporary Hungary. Social Compass 46(1):21–33.
Mead, Margaret
    1928   Coming of Age in Samoa. New York: Morrow.
Melchior, Marcus
    1965   A Rabbi Remembers. New York: Lyle Stuart.
Meyer, Poul
    1964   Gudstjenesten i brydningstiden efter 1814. In Ved 15-aars-dagen for
            Anordningen af 29. Marts 1814. J. Margolinsky and P. Meyer, eds. Pp.
            35–82. Copenhagen: Det Mosaiske Troessamfund.
Mitchell, Jon P.
    1998   The nostalgic construction of community: Memory and social identity
            in urban Malta. Ethnos 63(1):81–101.
Murphy, Joseph M.
    1995   Working the spirit: Ceremonies of the African diaspora. Boston: Beacon.
Myerhoff, Barbara
    1978   Number Our Days. New York: Simon and Schuster.
Nielsen, E. D.
    1955   N. F. S. Grundtvig: An American Study. Rock Island, NY: Augustana.
Nonini, Donald M.
    1999   The Dialectics of "Disputatiousness" and "Rice-eating Money": Class
            confrontation and gendered imaginaries among Chinese men in West
            Malaysia. American Ethnologist 26(1):47–68.

Oppenhejm, Mélanie
1981   Menneskefælden: Om livet i KZ-lejren Theresienstadt. Copenhagen: Hans Reitzel.
Oppenhejm, Ralph
1966   Det skulle så være : Dagbog fra Theresienstadt. Copenhagen: Carit Andersen.
Paulsson, Gunnar S.
1995   The "Bridge over the Øresund": The historiography on the expulsion of the Jews from Nazi-occupied Denmark. Journal of Contemporary History 30:431–464.
Pitt-Rivers, Julian
1954   The People of the Sierra. Chicago: University of Chicago Press.
Pitts, Walter.
1996   Old Ship of Zion: The Afro-Baptist Ritual in the African Diaspora. New York: Oxford University Press.
Pontoppidan Thyssen, Anders
1957   Den nygrundtvigske bevægelse med særligt henblik paa den Borupske kreds. Aarhus: Det Danske Forlag.
Radcliffe-Brown, A. R.
1948   The Andaman Islanders. Glencoe, Ill.: Free Press.
Raj, Dhooleka Sarhadi
2000   "Who the hell do you think you are?" Promoting religious identity among young Hindus in Britain. Ethnic and Racial Studies 23(3):535–558.
Redfield, Robert
1956   The Little Community and Peasant Society and Culture. Chicago: University of Chicago Press.
Rod, Jakob
1972   Dansk folkereligion i nyere tid. Copenhagen: G. E. C. Gad.
Rohde, Ina
1982   Da jeg blev jøde i Danmark. Copenhagen: Reitzel.
Roth, Nina
1999   Den polsk jødisk tvangsemigration 1969–1973: Et tilbageblik på tredive år i Danmark. RAMBAM 8(1):6–31.
Rottenberg, Dan, ed.
1997   Middletown Jews: The Tenuous Survival of an American Jewish Community. Bloomington: Indiana University Press.
Sachar, Abram Leon
1968   A History of the Jews. New York: Knopf.
Sacks, Jonathan
1994   From integration to survival to continuity: The third great era of modern Jewry. In Jewish Identities in the New Europe. J. Webber, ed. Pp. 107–116. London: Littman Library of Jewish Civilization.
Salamonsen, Per
1975   Religion i dag: Et sociologisk metodestudium. Copenhagen: G. E. C. Gad.

Sandemose, Aksel
   1936   A Refugee Crosses his Tracks. New York.
Schnapper, Dominique
   1987   Les limites de la démographie des juifs de la diaspora. Revue Français de Sociologie 28(2):319–332.
Scott, James
   1985   Weapons of the Weak. New Haven:Yale University Press.
Seeman, Don
   1999   "One People, One Blood": Public health, political violence, and HIV in an Ethiopian-Israeli setting. Culture, Medicine & Psychiatry 23(2):159–195.
Sered, Susan S.
   1988   Food and holiness: Cooking as a sacred act among middle-eastern Jewish women. Anthropological Quarterly 61(3):129–39.
Sered, Susan Starr
   1992   Women as Ritual Experts : The Religious Lives of Elderly Jewish Women in Jerusalem. New York: Oxford University Press.
Shokeid, Moshe, ed.
   1988   Children of Circumstances : Israeli Emigrants in New York. Ithaca: Cornell University Press.
——
   1997   Negotiating multiple viewpoints: The Cook, the Native, the Publisher, and the ethnographic text. Current Anthropology 38(4):631–645.
Shukla, Sandhya
   2001   Locations for South Asian diasporas. Annual Review of Anthropology 30(1):551–572.
Sode-Madsen, Hans
   1993   "Føreren har befalet!":Jødeaktionen oktober 1943. Copenhagen: Samleren.
——, ed.
   1995   Dengang i Theresienstadt : Deportationen af de danske jøder 1943–45. Copenhagen: Det Mosaiske Troessamfund.
Svendstorp, Aage
   1946   Den Hvide Brigade : Danske lægers modstand. Copenhagen: Aller Press.
Thomas, Nicholas
   1992   The inversion of tradition. American Ethnologist 19(2):213–232.
Torstrick, Rebecca L.
   2000   The Limits of Coexistence: Identity Politics in Israel. Ann Arbor: University of Michigan Press.
Trevor-Roper, Hugh
   1984   The invention of tradition:The highland tradition of Scotland. In The invention of Tradition. E. Hobsbawm and T. Ranger, eds. Pp. 15–41. Cambridge: Cambridge University Press.
Voye, L.
   1999   Secularization in a context of advanced modernity. Sociology of Religion 60(3):275–288.

Waters, Mary C.
1990   Ethnic Options: Choosing Identities in America. Berkeley: University of California Press.
Watson, James L., ed.
1997   Golden Arches East : McDonald's in East Asia. Stanford: Stanford University Press.
Webber, Jonathan, ed.
1994a   Jewish Identities in the New Europe. London: Littman Library of Jewish Civilization.
———
1994b   Modern Jewish identities. In Jewish Identities in the New Europe. J. Webber, ed. Pp. 74–85. London: Littman Library of Jewish Civilization.
———
1997   Jews and Judaism in contemporary Europe: Religion or ethnic group? Ethnic and Racial Studies 20(2):257–279.
Welner, Pinches
1965   Fra polsk jøde til dansk. Copenhagen: Steen Hasselbach.
Wolf, Eric
1982   Europe and the People Without History. Berkeley: University of California Press.
Yaari, Uri
1981   Indtryk fra Delegeretforsamlingen. In Jodisk Orientering. Pp. 9–10, Vol. 52.
Yahil, Leni
1969   The Rescue of Danish Jewry: Test of a Democracy. Philadelphia: Jewish Publication Society of America.
Yelvington, Kevin A.
2001   The anthropology of Afro-Latin America and the Caribbean: Diasporic dimensions. Annual Review of Anthropology 30(1):227–260.
Zborowski, Mark, and Elizabeth Herzog
1952   Life Is with People: The Culture of the Shtetl. New York: Schocken.
Zenner, Walter, ed.
1988   Persistence and Flexibility: Anthropological Perspectives on the American Jewish Experience. Albany: State University of New York Press.
———
1991   Minorities in the Middle: A Cross-cultural Analysis. Albany: State University of New York Press.
Zenner, Walter P.
1985   Jewishness in America: ascription and choice. Ethnic and Racial Studies 8(1):117–133.
Zerubavel, Yael
1994   Recovered roots: Collective Memory and the Making of Israeli National Tradition. Chicago: University of Chicago Press.
Zuckerman, Phil
1999   Strife in the Sanctuary: Religious Schism in a Jewish Community. Walnut Creek: Altamira Press.

Index